Usability Evaluation for In-Vehicle Systems

Usability Evaluation for In-Vehicle Systems

Catherine Harvey
Neville A. Stanton

CRC Press is an imprint of the
Taylor & Francis Group, an **informa** business

Cover designed by Josh Stanton.

CRC Press
Taylor & Francis Group
6000 Broken Sound Parkway NW, Suite 300
Boca Raton, FL 33487-2742

© 2013 by Taylor & Francis Group, LLC
CRC Press is an imprint of Taylor & Francis Group, an Informa business

No claim to original U.S. Government works

Printed on acid-free paper
Version Date: 20130222

International Standard Book Number-13: 978-1-4665-1429-4 (Hardback)

This book contains information obtained from authentic and highly regarded sources. Reasonable efforts have been made to publish reliable data and information, but the author and publisher cannot assume responsibility for the validity of all materials or the consequences of their use. The authors and publishers have attempted to trace the copyright holders of all material reproduced in this publication and apologize to copyright holders if permission to publish in this form has not been obtained. If any copyright material has not been acknowledged please write and let us know so we may rectify in any future reprint.

Except as permitted under U.S. Copyright Law, no part of this book may be reprinted, reproduced, transmitted, or utilized in any form by any electronic, mechanical, or other means, now known or hereafter invented, including photocopying, microfilming, and recording, or in any information storage or retrieval system, without written permission from the publishers.

For permission to photocopy or use material electronically from this work, please access www.copyright.com (http://www.copyright.com/) or contact the Copyright Clearance Center, Inc. (CCC), 222 Rosewood Drive, Danvers, MA 01923, 978-750-8400. CCC is a not-for-profit organization that provides licenses and registration for a variety of users. For organizations that have been granted a photocopy license by the CCC, a separate system of payment has been arranged.

Trademark Notice: Product or corporate names may be trademarks or registered trademarks, and are used only for identification and explanation without intent to infringe.

Library of Congress Cataloging-in-Publication Data

Harvey, Catherine, 1984-
  Usability evaluation for in-vehicle systems / Catherine Harvey, Neville A. Stanton.
    pages cm
  Includes bibliographical references and index.
  ISBN 978-1-4665-1429-4 (hardback)
  1. Automobiles--Instruments--Display systems. 2. Human-machine systems. I. Stanton, Neville, 1960- II. Title.

TL272.55.H37 2013
629.2'73--dc23   2013003951

Visit the Taylor & Francis Web site at
http://www.taylorandfrancis.com

and the CRC Press Web site at
http://www.crcpress.com

Contents

Preface xi
Acknowledgements xiii
The Authors xv
Glossary xvii

Chapter 1 Introduction 1

    The History of In-Vehicle Information Provision 1
        Instrumentation 3
        Infotainment 5
        Navigation 7
        Comfort 8
        Future Predictions 8
    Ergonomics Challenges of In-Vehicle Information Systems (IVIS) ... 10
    Ergonomics, Human Computer Interaction (HCI), and Usability 12
    Usability Evaluation 14
    Book Outline 15

Chapter 2 Context-of-Use as a Factor in Determining the Usability of
In-Vehicle Information Systems 19

    Introduction 19
    Significant Contributions to Defining Usability 20
        Brian Shackel 20
        Jakob Nielsen 21
        Donald Norman 22
        Ben Shneiderman 23
        Nigel Bevan and The International Organization for
        Standardization (ISO) 23
    The Development of a Definition of Usability 26
        A Universal Definition of Usability? 26
        Specifying Usability Factors 28
    Usability of In-Vehicle Information Systems (IVIS) 29
        Defining the Context-of-Use 29
            Dual Task Environment 30
            Environmental Conditions 31
            Range of Users 32
            Training Provision 32
            Frequency of Use 33
            Uptake 33
        Defining Usability Factors for IVIS 34
    Conclusions 36

Chapter 3 In-Vehicle Information Systems to Meet the Needs of Drivers 37

Introduction 37
The Task 38
  Primary Driving Tasks 38
  Secondary (In-Vehicle) Tasks 39
The System 39
  Touch Screens and Remote Controllers 40
The User 41
  Safety 42
  Efficiency 43
  Enjoyment 43
The Task-User-System Interaction 44
  Multimodal Interactions 44
  Toward a Prediction of IVIS Usability 45
Evaluating the Task–System–User Interaction 45
Conclusions 47

Chapter 4 A Usability Evaluation Framework for In-Vehicle Information Systems 49

Introduction 49
Preparing for a Usability Evaluation 49
Selecting Usability Evaluation Methods 49
  Information Requirements for Usability Evaluations 50
    Objective Measures 51
    Subjective Measures 51
  When to Apply Methods in IVIS Usability Evaluations 52
    Analytic Methods 52
    Empirical Methods 52
  Resources Available for IVIS Usability Evaluations 52
    Representing the System and Tasks 52
    Representing the User 53
    The Testing Environment 53
  People Involved in IVIS Usability Evaluations 53
    Usability Evaluation with Users 54
    Usability Evaluation with Experts 54
  A Flowchart for Method Selection 54
Usability Evaluation Methods 55
  Analytic Evaluation Methods 57
    Hierarchical Task Analysis (HTA) 57
    Multimodal Critical Path Analysis (CPA) 58
    Systematic Human Error Reduction and Prediction
     Approach (SHERPA) 58
    Heuristic Analysis 59
    Layout Analysis 59

Contents   vii

                  Empirical Evaluation Methods 60
                      Objective Methods 60
                      Subjective Methods 63
                Conclusions 64

Chapter 5   The Trade-Off between Context and Objectivity in an Analytic Evaluation of In-Vehicle Interfaces 71

                Introduction 71
                Analytic Methods 72
                Method 72
                    Equipment.. 72
                      The IVIS 72
                      Data Collection Apparatus 74
                  Procedure... 74
                  Data Analysis.. 76
                Results and Discussion 77
                  Hierarchical Task Analysis (HTA).................................... 77
                      HTA for IVIS Evaluation....................................... 77
                      HTA Utility.. 79
                  Multimodal Critical Path Analysis (CPA) 80
                      Defining CPA Activities 80
                      CPA for IVIS Evaluation 83
                      CPA Utility 87
                  Systematic Human Error Reduction and Prediction Approach (SHERPA)... 87
                      SHERPA for IVIS Evaluation............................... 87
                      SHERPA Utility... 92
                  Heuristic Analysis 92
                      Heuristic Analysis for IVIS Evaluation 92
                      Heuristic Analysis Utility 92
                  Layout Analysis 94
                      Layout Analysis for IVIS Evaluation.................... 94
                      Layout Analysis Utility 96
                General Discussion... 96
                  Analytic Methods for IVIS Evaluation 97
                  Context versus Objectivity.. 100
                  Extending CPA 100
                Conclusions 101

Chapter 6   To Twist or Poke? A Method for Identifying Usability Issues with Direct and Indirect Input Devices for Control of In-Vehicle Information Systems 103

                Introduction 103
                Direct and Indirect IVIS Input Devices 104

Empirical Evaluation of IVIS Usability 105
Selection of Tasks 105
   Types of Operation 107
Method 108
   Participants 108
   Equipment 108
      The University of Southampton's Driving Simulator 108
      In-Vehicle Information Systems 109
      Eye Tracking 110
      User Questionnaires 110
   Procedure 112
      Secondary In-Vehicle Tasks 112
   Data Collection and Analysis 113
Results and Discussion 114
   Primary Driving Performance 114
      Longitudinal Control 115
      Lateral Control 116
   Visual Behaviour 117
   Secondary Task Performance 119
      Secondary Task Times 119
      Secondary Task Errors 121
   Subjective Measures 122
      System Usability Scale (SUS) 122
      Driving Activity Load Index (DALI) 123
   Usability Issues 126
      Rotary Controller Usability Issues 127
      Touch Screen Usability Issues 128
      Graphical User Interface/Menu Structure Usability Issues 128
      Optimisation between the Graphical User Interface (GUI),
      Task Structure, and Input Device 128
   Implications 129
Conclusions 130

Chapter 7 Modelling the Hare and the Tortoise: Predicting IVIS Task Times for Fast, Middle, and Slow Person Performance using Critical Path Analysis 133

Introduction 133
Modelling Human–Computer Interaction 133
   Task Times 134
   Modelling Techniques 134
Critical Path Analysis 136
   Extending CPA for Fastperson and Slowperson Predictions 137
Method 137
Identification of Operation Times 137

Contents

	Development of the CPA Calculator	142
	Comparison of CPA-Predicted Task Times with Empirical Data	143
	Participants	143
	Equipment	144
	Procedure	144
	Data Collection and Analysis	144
	Results	145
	Discussion	147
	Applications of the CPA Model	148
	Limitations of the CPA Model	150
	Extensions to the CPA Model	151
	Conclusions	152
Chapter 8	Visual Attention on the Move: There Is More to Modelling than Meets the Eye	153
	Introduction	153
	The CPA Method	153
	Visual Behaviour in Driving	154
	Dual-Task Glance Durations: Previous Findings	154
	Method	155
	Glance Behaviour Data Analysis	155
	Development of a CPA Model for Dual-Task IVIS Interaction	156
	Glance Behaviour Data	156
	Model Assumptions	156
	Dual-Task CPA Calculator	158
	Results	158
	Case Study: Glance Behaviour in a Dual-Task Environment	159
	Shared Visual Attention	159
	Glance Behaviour for Sequential Operations	163
	Results: Shared Glance CPA Model	165
	Discussion	167
	Multidimensionality	167
	Implications for Visual Behaviour Theory in Driving	168
	The Occlusion Technique	169
	Road and IVIS Glance Durations	169
	Implications for IVIS Design	171
	Limitations of the CPA Model	171
	Conclusions	172
Chapter 9	Summary of Contributions and Future Challenges	175
	Introduction	175
	Summary of the Findings	175
	Novel Contributions of the Work	177

Key Questions	179
Areas for Future Research	182
Concluding Remarks	185
References	187
Index	207

Preface

The work presented in this book was prompted by the need for an evaluation framework that is useful and relevant to the automotive industry. It is often argued that Ergonomics is involved too late in commercial project development processes to have substantive impact on design and usability. In the automotive industry, and specifically in relation to In-Vehicle Information Systems (IVIS), a lack of attention to the issue of usability not only can lead to poor customer satisfaction but can also present a significant risk to safe and efficient driving. This work contributes to the understanding and evaluation of usability in the context of IVIS and is written for students, researchers, designers, and engineers who are involved or interested in the design and evaluation of in-vehicle systems. The book has three key objectives

- Define and understand usability in the context of IVIS. This guides the specification of criteria against which usability can be successfully evaluated.
- Develop a multimethod framework to support designers in the evaluation of IVIS usability. The underlying motivations for the framework are a need for early-stage evaluation to support proactive redesign and a practical and realistic approach that can be used successfully by automotive manufacturers.
- Develop an analytic usability evaluation method that enables useful predictions of task interaction, while accounting for the specific context-of-use of IVIS. The major challenge of this particular context-of-use is the dual-task environment created by interacting with secondary tasks via an IVIS at the same time as driving.

In order to meet these objectives, within the book we have examined how usability evaluation of IVIS can help designers to understand the limitations of current systems in order to develop new concept technologies. The aim of the work is to further readers' understanding of how they can develop more usable systems to enhance the overall driving experience by meeting the needs of the driver for safety, efficiency, and enjoyment. This book is aimed at designers and engineers involved in the development of in-vehicle systems, researchers, and students within the disciplines of Human–Computer Interaction, Ergonomics, and Psychology, and Ergonomics practitioners. We hope that those working in the practical development of in-vehicle systems will make use of the various methods described here in their work on the development and evaluation of future products and will also benefit from the insights into both the theory and empirical findings presented in the book. At the same time, we expect that researchers and students will find the theoretical concepts useful and interesting and that the hypotheses put forward in the book will stimulate further work in this area.

Acknowledgements

We would like to thank Jaguar Land Rover and the Engineering and Physical Sciences Research Council (EPSRC) for providing the research funding that supported this work.

Special thanks go to Prof. Mike McDonald, Dr. Pengjun Zheng, Dr. Carl Pickering, and Lee Scrypchuk for supporting this work and for much helpful advice throughout the project. We would also like to thank Tom Jones and Alec James, and everyone who participated in the simulator studies for their invaluable contribution to the empirical phases of this work.

PERSONAL ACKNOWLEDGEMENTS

For Alan

Cath

For Maggie, Josh, and Jem

Neville

The Authors

Catherine Harvey
Transportation Research Group, Faculty of Engineering and the Environment, University of Southampton, Highfield, Southampton, SO17 1BJ, UK
c.harvey@soton.ac.uk

Dr. Catherine Harvey is a research fellow within the Transportation Research Group at the University of Southampton, UK. Catherine was awarded an engineering doctorate (EngD) in 2012 from the University of Southampton. Her research focussed on the usability of in-vehicle interfaces and Ergonomics evaluation methods, and was sponsored by Jaguar Land Rover. Catherine also has a BSc in Ergonomics (Hons) from Loughborough University. She is currently working on the EU Seventh Framework funded project ALICIA (All Condition Operations and Innovative Cockpit Infrastructure), focussing on the evaluation of flight deck display concepts and the pilot–flight deck interaction.

Neville A. Stanton
Transportation Research Group, Faculty of Engineering and the Environment, University of Southampton, Highfield, Southampton, SO17 1BJ, UK
n.stanton@soton.ac.uk

Prof. Neville Stanton holds the chair in human factors engineering at the University of Southampton. He has published more than 170 peer-reviewed journal papers and 25 books on Human Factors and Ergonomics. In 1998, he was awarded the Institution of Electrical Engineers Divisional Premium Award for work on Engineering Psychology and System Safety. The Institute of Ergonomics and Human Factors awarded him the Otto Edholm medal in 2001, the President's Medal in 2008 and the Sir Frederic Bartlett Medal in 2012 for his original contribution to basic and applied Ergonomics research. In 2007, The Royal Aeronautical Society awarded him the Hodgson Medal and Bronze Award with colleagues for their work on flight deck safety. Prof. Stanton is an editor of the journal *Ergonomics* and on the editorial boards of *Theoretical Issues in Ergonomics Science* and the journal of *Human Factors and Ergonomics in Manufacturing and Service Industries*. Prof. Stanton consults for a wide variety of organisations on topics such as human factors, safety cases, safety culture, risk assessment, human error, product design, warning design, system design, and operation. He has also acted as an expert witness. He is a fellow and Chartered Occupational Psychologist registered with the British Psychological Society, and a Fellow of the Ergonomics Society. He has a BSc (Hons) in Occupational Psychology from the University of Hull, an MPhil in Applied Psychology and a PhD in human factors from Aston University in Birmingham.

Glossary

ACT-R: Adaptive Control of Thought-Rational
ANOVA: Analysis of Variance
CHI: Conference on Human Factors in Computing Systems
CPA: Critical Path Analysis
CPM-GOMS: Cognitive Perceptual Motor—Goals, Operators, Methods and Selection rules
DALI: Driving Activity Load Index
DETR: Department of the Environment, Transport and the Regions
DfT: Department for Transport
EFT: Early Finish Time
EngD: Engineering Doctorate
EPIC: Executive Process—Interactive Control
EST: Early Start Time
GOMS: Goals, Operators, Methods and Selection rules
GPS: Global Positioning System
GSM: Global System for Mobile communications
GUI: Graphical User Interface
HCI: Human–Computer Interaction
HMI: Human–Machine Interface
HTA: Hierarchical Task Analysis
HUD: Head-Up Display
IQR: InterQuartile Range
ISO: International Organization for Standardization
IVIS: In-Vehicle Information System
KLM: Keystroke Level Model
KPI: Key Performance Indicator
LCD: Liquid Crystal Display
LEAF: Learnability, Effectiveness, Attitude, and Flexibility
LFT: Late Finish Time
LST: Late Start Time
MHP: Model Human Processor
MMI: (Audi) MultiMedia Interface
MPH: Miles Per Hour
ms: Milliseconds
NASA-TLX: National Aeronautics and Space Administration—Task Load Index
NMT: Nordic Mobile Telephone network
OEM: Original Equipment Manufacturer
s: Seconds
SD: Standard Deviation
SHERPA: Systematic Human Error Reduction and Prediction Approach
SUS: System Usability Scale

TRL: Transport Research Laboratory
UFOV: Useful Field of View
VCR: Video Cassette Recorder
VDT: Visual Display Terminal
VDU: Visual Display Unit

1 Introduction

THE HISTORY OF IN-VEHICLE INFORMATION PROVISION

From the very early days of the motor car it has been essential for drivers to be able to quickly and easily operate all of the vehicle controls, as well as know the state of the vehicle, which in its most basic form consists of information about speed and fuel level (Damiani et al., 2009). Ergonomics has played a major role in the development of the driver–vehicle interface and evidence for consideration of ease of use can be found as early as 1907, in *A Busy Man's Textbook on Automobiles* (Faurote, 1907):

> Everything has been designed with an eye to accessibility. In all of the four cylinder cars the motor is placed forward under a sheet metal hood in such a position that every part is quickly 'get-a-ble'.... In fact, taken all in all, the general trend of motor car design is to make a machine which will be practical, comfortable and serviceable. (p. 28)

Factors such as accessibility, which are now recognised as central to the Ergonomics purview, were initially considered in relation to the mechanics of the car; however, the need for feedback and information in the vehicle and the vast developments in display technologies over the last century have resulted in dramatic changes to the vehicle's interior. These changes are notable in the design of the dashboard and instrument cluster, and the focus of Ergonomics has shifted accordingly. The introduction of dashboard controls prompted concerns about the driver's physical comfort, particularly as the motor car became more accessible to a wider range of users. An example is given in Figure 1.1, which shows an advertisement from 1969 by the automotive manufacturer General Motors, promoting their appointment of a female 'stylist' to consider the physical reach of women vehicle occupants. Although the approach to vehicle Ergonomics may have evolved somewhat since the early days of the motor vehicle, it is clear that issues such as accessibility and usability have been pertinent for almost as long as people have been driving cars. There is more information available in contemporary vehicles than ever before, which places significant importance on the role of not just physical but also cognitive Ergonomics in vehicle design. In fact, technical evolution has occurred at such a pace that there is potential for it to exceed the capabilities of the driver (Rumar, 1993; Walker et al., 2001; Peacock and Karwowski, 1993) if the in-vehicle interface and the needs of the driver are given insufficient focus during the product development process.

In order to provide some historical context for the book, this chapter documents the development of in-vehicle display and interaction technologies, starting with the first speedometer, introduced in the early 1900s, and finishing with some predictions

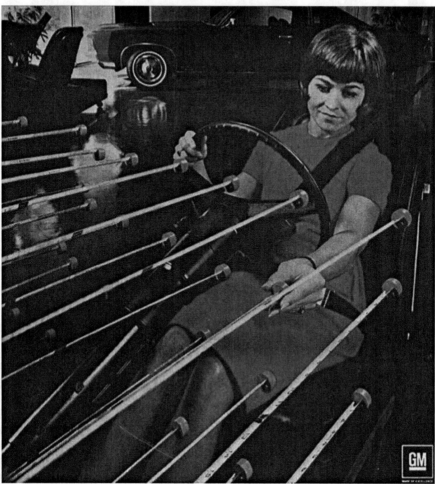

FIGURE 1.1 A 1969 advertisement by General Motors highlighting the importance of females' physical reach in vehicle interior design. (Image reproduced courtesy of General Motors.)

Introduction

for future technological developments within the vehicle. Four timelines were constructed to document the development of in-vehicle systems instrumentation, infotainment, navigation, and comfort; these are each presented alongside a discussion of the technologies. The timelines were based on information from a number of sources, including Azuma et al. (1994), Damiani et al. (2009), Ludvigsen (1997), Newcomb and Spurr (1989), and the websites of individual automotive manufacturers and aftermarket equipment manufacturers.

INSTRUMENTATION

Instrumentation describes the Human–Machine interface (HMI) that connects the driver and the vehicle from the clocks and dials of early motor cars to the digital multi-menu-level displays of today's automobiles (see Figure 1.2). One of the earliest forms of instrumentation in the motor car was a simple speedometer (Newcomb and Spurr, 1989), which was first developed around the turn of the twentieth century by Otto Schulze. Basic instrumentation was attached to the dashboard, which in early motor cars was fixed behind the engine hood to prevent pebbles being dashed from the roadway onto the vehicle occupants. This meant that the instruments were located very low in the vehicle, barely visible to the driver (Ludvigsen, 1997). This lack of accessibility to controls was overcome with the introduction of the instrument panel and integration of instrumentation and controls for the radio and climate appearing between the 1930s and '50s (Ludvigsen, 1997; Newcomb and Spurr, 1989). The provision of a dedicated instrument panel also meant that the driver had full control over the increasing variety of functions appearing in cars during the mid- to late 20th century, including the radio, passenger comfort, and navigation. These in-vehicle functions were controlled by instrument panel dials for many years until

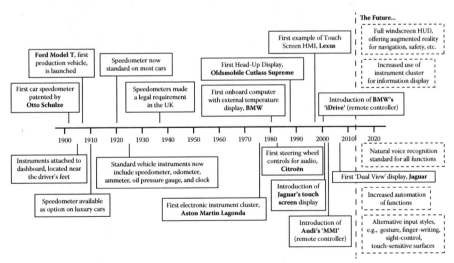

FIGURE 1.2 In-vehicle instrumentation: input and output trends.

FIGURE 1.3 BMW's iDrive IVIS. (Author's own photographs.)

the development and widespread introduction of In-Vehicle Information Systems (IVIS), which integrated the controls for many in-vehicle functions into a single screen-based device. One of the earliest examples of this new vehicle interaction style was Lexus's touch screen system, introduced in 1997. This was closely followed by other touch screen devices (e.g., Jaguar) and by an alternative remote control interaction style, the most well-known being BMW's iDrive system, introduced in 2001 (see Figure 1.3). These IVIS changed the appearance of traditional vehicle dashboards by reducing the number of separate hard dials. This development also increased the capacity for functionality in the vehicle, and today's car offers infinitely more infotainment, navigation, communication, and comfort options than the early motor vehicle. A recent example of a touch screen IVIS from Jaguar Cars is shown in Figure 1.4; this displays a single screen-based interface located in the centre console, which integrates many secondary vehicle functions into a single system. Further advances in display technology for the vehicle have included the Head-Up Display (HUD), which was first introduced in cars in 1988 by the vehicle manufacturer Oldsmobile and is predominantly used for displaying basic information such as speed, parking monitor displays (including the first 360-degree Around-View system by Infiniti [2010]), and 'dual view' screens, introduced in the vehicle by Jaguar in 2007 (Jaguar Cars Limited, 2011).

Introduction 5

FIGURE 1.4 An example of a modern touch screen IVIS as seen in Jaguar vehicles. (Image reproduced courtesy of Jaguar Cars.)

INFOTAINMENT

Infotainment refers to the provision of information, entertainment, and communication functions within the vehicle; the development timeline is shown in Figure 1.5. The earliest attempts at providing radio to the driver consisted of a radio receiver connected to an amplifier and loud speaker, which was mounted on the side of a vehicle. An example, the Marconiphone, is shown connected to the running board of a car in Figure 1.6. Radio was the first source of entertainment to be integrated into the vehicle, with one of the earliest examples being the Galvin Manufacturing Company's Motorola in 1930, which was specifically designed as a car radio (Motorola Solutions, 2012). This proved a popular addition, and shortly afterwards many vehicle manufacturers began to fit radio wiring in their vehicles as standard (Ludvigsen, 1997). Building on the popularity of the in-car radio, manufacturers started to introduce recorded media into the vehicle, beginning with 45-rpm records, although these were unable to successfully withstand the vibrations from the road surface (Ludvigsen, 1997), and were soon replaced when Philips produced the first standard cassette player in the late 1960s (Koninklijke Philips Electronics, 2012). In-car entertainment kept pace with developments in prerecorded media, with the first vehicle CD player developed in the mid-80s by Pioneer (Pioneer Europe, 2012), followed by MP3 capability in the late 1990s (Empeg, 2012). More recently, video capabilities have been added to the vehicle, with the first in-car Blu-ray players introduced within the last 5 years. The addition of visual entertainment could be perceived as a worrying trend in terms of distraction and driver safety; however, manufacturers have gone some way to acknowledge these issues, introducing technologies such as Jaguar's Dual-View screen, which displays different images depending on the viewing angle, thereby restricting the view of moving images to the passenger only (Jaguar Cars Limited, 2011).

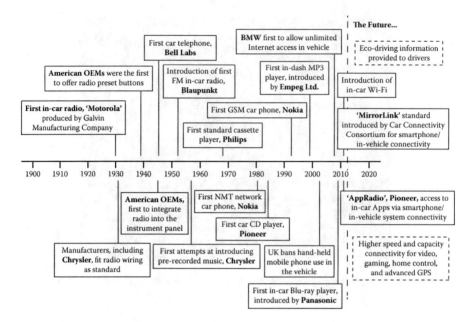

FIGURE 1.5 In-vehicle infotainment trends.

FIGURE 1.6 Marconiphone V2 wireless receiver, mounted on the running board of a car. (Image reproduced courtesy of Telent Limited.)

Introduction

Another aspect of infotainment functionality, communication, was an important feature in the development of the motor car, with the first car telephone pioneered by Bell Labs in 1946 (AT&T, 2012). Further improvements in car phone technology was driven by the development of mobile communication networks: Nokia introduced the first car phones for the first fully automatic cell phone system, the Nordic Mobile Telephone (NMT) network, established in 1981 (Nokia, 2009; Staunstrup, 2012). This was followed by the Global System for Mobile Communications (GSM) network, created in the early 1990s, which was accompanied by the first GSM car phone, introduced by Nokia in 1993 (Nokia, 2009). Still keeping relative pace with the technological development of communication technologies, automotive manufacturers allowed Internet access in the car in the late 2000s. Today, many drivers are able to access the web via a smartphone connected to the vehicle, and there are even 'apps' designed specifically for in-vehicle use.

NAVIGATION

Vehicle navigation was also an area of much interest from the very early days of motor vehicle (see Figure 1.7). The Jones Live-Map, shown in Figure 1.8, is the earliest example of a navigation system. Developed in 1909, it used a cable connecting the car's front wheels to turn a circular map indicating the vehicle's position along a route (Moran, 2009). Development in this area, however, did not really flourish until the 1980s, when Honda introduced the Electro Gyrocator, which was the first gyroscopic navigation system, incorporating a mechanism which scrolled static maps through a display to indicate position (Honda Motor Company, 2012). The first Global Positioning System (GPS) navigation device, Toyota's Crown system, was introduced in 1988 (Azuma et al., 1994), followed shortly after by the first aftermarket navigation system, developed by Pioneer in 1991 (Pioneer Europe, 2012).

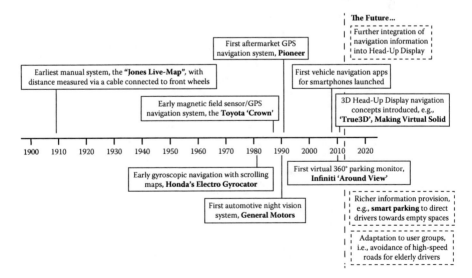

FIGURE 1.7 In-vehicle navigation system trends.

FIGURE 1.8 The Jones Live-Map, one of the earliest automobile navigation systems. (Image reproduced courtesy of Skinner Inc., www.skinnerinc.com.)

More recently, there has been a huge increase in sales of aftermarket, or nomadic, navigation devices (Stevens, 2012), most offering a touch-screen interface; however, with the continuous improvements in smartphone technology, for instance, screen size and resolution, and the increasing availability of navigation apps tailored to the car, it is likely that aftermarket navigation-only devices will experience a decline in popularity over the coming years (Stevens, 2012). Supplementary navigation-related functions also continue to emerge, and recent years have seen the introduction of night-vision systems, parking monitors, and 3D route displays integrated into a HUD.

COMFORT

The early automobile was open to the elements, a feature which was, in fact, seen to increase the pleasure of driving a car (Ludvigsen, 1997). In these very early days, the driver and passengers had to adjust themselves to the car (Ludvigsen,1997); however, as the motor car developed, more attention was paid to comfort, initially of rear-seat occupants, followed later by the driver. The first front- and rear-compartment heating systems began to appear in the early twentieth century (see Figure 1.9), with the first heating system developed by Ford in 1933 and the first air conditioning system produced by the Packard Motor Company in 1939 (Ludvigsen, 1997). Modern automobiles now adjust themselves to the driver, providing heating and cooling to different parts of the vehicle, automatically controlled or activated by the driver via the IVIS or dashboard dials.

FUTURE PREDICTIONS

Based on the rate of recent advances in in-vehicle technology it is very likely that development will continue at pace with the introduction of many more functions and devices into the vehicle. In an attempt to forecast some of the advances in IVIS

Introduction

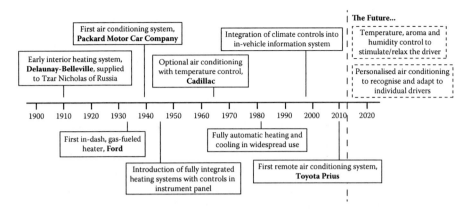

FIGURE 1.9 In-vehicle comfort control trends.

expected over the next decade, predictions for future in-vehicle technology have been incorporated into the timelines. In terms of instrumentation and interaction styles, it is likely that the expected improvements in technology will allow natural voice recognition to become an important component of vehicle feedback and control in the near future. Other alternative input styles are also likely to be seen more widely in cars, including gesture recognition, full HUD, and augmented reality applications. Cars move increasingly toward a 'glass cockpit/dashboard' concept as predicted by Walker et al. (2001), with more and more instruments moving to digital displays and the opportunities for driver-customization of displays looking likely. Automation is currently a very popular topic in vehicle research (see, for example, Heide and Henning, 2006; Jamson et al., 2011; Khan et al., 2012; Stanton et al., 2007b; Stanton et al., 2011; Rudin-Brown, 2010; Young and Stanton, 2007), and increasing levels of automation may actually reduce the number of functions with which the driver needs to interact. However, this is not expected to reduce the variety of interaction styles introduced by original equipment manufacturers (OEMs), as there appears to be an ever-intensifying drive toward novel interaction technologies. Over the last decade the need for adaptive information presentation has been acknowledged by researchers and automotive manufacturers (see Amditis et al., 2006; Hoedemaeker and Neerincx, 2007; Piechulla et al., 2003; Rudin-Brown, 2010; Sarter, 2006; Walker et al., 2001). Adaptive in-vehicle systems assess the state of the driver (e.g., awareness and fatigue) and provide real-time information or warnings to assist the driver or restrict access to particular functions in situations of high workload. The first examples of adaptive information presentation are now being seen in production vehicles, with Lexus introducing the world's first Advanced Pre-Collision System, which uses an infrared camera to monitor the direction of the driver's face. If the driver is facing away, and the system detects a potential obstacle in the vehicle's path, the car can automatically apply the brakes and retract the front seatbelts (Lexus, 2012). Other vehicle manufacturers, including Volkswagen and Daimler, have recently introduced fatigue monitoring systems that produce auditory and visual alerts when drowsiness is detected (Daimler, 2012; Volkswagen, 2012). There is also currently much attention on eco-driving (Birrell et al., 2011; Damiani et al., 2009; Flemming et al.,

2008; Young et al., 2011b), and with more hybrid and electric vehicle entering the market, in-vehicle information provision will need to account for differences in driving styles and vehicle operation brought about by these developments (Young et al., 2011b). Focus on the comfort of the driver is likely to continue (Damiani et al., 2009), and technologies that enable the car to adapt the internal thermal environment to the current state of the driver are expected in the near future. Increasing levels of connectivity, via high-speed, wide-ranging Internet access and the use of portable devices such as smartphones, means that vehicle users now expect to be able to connect to a massive amount of information whilst driving (Damiani et al., 2009). Recent initiatives such as MirrorLink (Car Connectivity Consortium, 2011), which is an open-industry standard for car-centric connectivity solutions, seek to increase the connectivity between portable devices and in-vehicle systems, and there is likely to be an increase in the number of in-vehicle devices that offer this seamless integration to people's smartphones and personal computers. This proliferation of in-vehicle technology is intended to enhance the driving experience; however, it is not without its disadvantages, as the higher the demand for attention inside the vehicle, the less attention is available for eyes-out monitoring of the road. This poses a considerable Ergonomics challenge for automotive manufacturers and researchers into the future.

ERGONOMICS CHALLENGES OF IN-VEHICLE INFORMATION SYSTEMS (IVIS)

Over the last decade, IVIS have become established as a standard technology in many road vehicles. Since the introduction of these multifunctional, menu-based systems in vehicles around the beginning of the 21st century, they have attracted much attention, and this has not always been positive. This has brought the concept of usability into sharp focus. Ten years ago the main focus of attention was on how much technology could be brought into vehicles. Today, the challenge is balancing the ever-increasing demand for technology with the users' needs, not only for form and function, but also for a usable HMI.

In 2011 there were 203,950 reported road casualties in Great Britain, although the Department for Transport (DfT) estimated the actual number to be nearer to seven hundred thousand every year as many accidents, particularly those involving non-fatal casualties, are not reported to the police (Department for Transport, 2012). Distraction in the vehicle was a contributory factor in almost three thousand of the reported road accidents in 2010 (Department for Transport, 2011a). This amounted to two per cent of all reported accidents; however, the World Health Organization (2011) suggested that this is likely to be an underestimate because of the difficulty in identifying distraction related incidents. In the United States, 18% of injury crashes in 2010 were described as distraction-affected (National Highway Traffic Safety Administration, 2012). Cars are now constructed to make driving safer than ever, but the risk from performing secondary tasks within the vehicle remains a significant threat to driver safety (Regan et al., 2009; Young et al., 2008). Secondary driving tasks are not directly involved in driving (Hedlund et al., 2006) and relate to the control of infotainment, comfort, navigation and communication functions. Primary driving tasks include steering, braking, controlling speed, manoeuvring in traffic, navigating to a destination and scanning for hazards (Hedlund et al., 2006), with the aim of maintaining safe control of the vehicle (Lansdown, 2000). Interaction with secondary tasks is a potential cause of in-vehicle distractions because it can increase the demands on the driver's visual, cognitive, auditory, and physical resources and this may result in a reduction in the driver's attention to the primary driving task (Burnett and Porter, 2001; GuJi and Jin, 2010; Hedlund et al., 2006; Lee et al., 2009; Matthews et al., 2001; Young and Stanton, 2002; Hancock et al., 2009b).

Traditionally, secondary functions were operated via a series of hard switches mounted on the vehicle's dashboard; see Figure 1.10. Today, in the premium sector, and increasingly with volume brands, these functions are integrated into a single menu-based system, for example, the rotary controller (Figure 1.3) and the touch screen (Figure 1.4), with only the most high-frequency and high-importance controls left as hard switches. The IVIS make use of a screen-based interface, which reduces the cluttered appearance of the dashboard and is considered to be an aesthetically superior solution to the traditional layout (Fleischmann, 2007). The ease with which a driver can interact with an IVIS is determined by the HMI because this influences a driver's ability to input information to the IVIS, to receive and understand information outputs, and to monitor the state of the system. As a result of the demand for enhanced in-vehicle functionality, IVIS complexity is increasing at a rate which is, in some cases, exceeding human capabilities: this is likely to result in an increase in driver distraction (Walker et al., 2001). This problem was exemplified by the BMW iDrive, released in 2000 (see Figure 1.3). Despite other similar systems by Audi and Mercedes to name but a few coming under fire for lack of efficiency and excessive menu complexity (Cunningham, 2007), it was BMW's rotary-controlled IVIS, which received the most high-profile criticism from the media and users alike (Cobb, 2002; Farago, 2002). Accusations that the iDrive lacked learnability (Cunningham, 2007), introduced redundancy on the dashboard, and attempted to incorporate vastly more functions into the vehicle than the driver would ever need (Cobb, 2002) pushed the issue of usability to the wider attention of the public. As the iDrive attracted increasing notoriety, reports like that by Farago (2002), which claimed that the iDrive was not 'a new way to drive' as BMW intended but rather a 'new way to die', served to fuel the attention on the issues of driver distraction and safety. This prompted much research into the interaction between the driver and IVIS and the effects of this on distraction (for an overview see Beirness et al., 2002; Carsten, 2004; Lansdown et al., 2004a, Lee et al., 2009; Young et al., 2003). With the ever-evolving face of IVIS, the transforming role of the driver from operator to monitor with increasing automation and the changes in driver demographics expected over the next decades, this research is still in full flow.

Increased complexity of IVIS interactions has been shown to be linked to poor driving performance; for example, Horrey (2011) reported that more complex tasks tend to result in longer glances away from the road than easier tasks, resulting in a lack of awareness of the road environment. This illustrates a situation in which the demands of an IVIS task exceed the capabilities of the driver, resulting in the degradation of the driver's visual attention to the road. The design of new in-vehicle technologies must account for this mismatch between IVIS complexity and the driver's

FIGURE 1.10 Traditional switch-based layout of the vehicle dashboard. (Author's own photograph.)

capabilities; otherwise, the benefits offered by the growth in in-vehicle functionality will be outweighed by the associated rise in distraction and consequent risk to safety (Hedlund et al., 2006; Hancock et al., 2009b). In its Strategic Framework for Road Safety, the DfT identified the potential for new technology to cause driver distraction as an important factor for the future of road safety (Department for Transport, 2011b). The DfT acknowledged that whilst the continued development of in-vehicle technologies is expected, there is a need to encourage manufacturers towards a solution that enables these technologies to be used safely within the car (Department for Transport, 2011b). This book aims to directly address these issues.

ERGONOMICS, HUMAN COMPUTER INTERACTION (HCI), AND USABILITY

Ergonomics can be defined as 'the application of scientific information concerning humans to the design of objects, systems, and environments for human use' (Whitfield and Langford, 2001). Ergonomics is 'an applied, multidisciplinary subject' (Buckle, 2011) that uses analysis and understanding to optimize design, 'augmenting and supporting human endeavour' (Lindgaard, 2009). Above all, Ergonomics focusses on identifying the needs of the user and designing to address these needs (Caple, 2010). The application of Ergonomics transcends disciplines, making products more usable, workplaces safer and transport systems more efficient. It can contribute to solving today's big issues, such as supporting our aging population and reducing the risk of major infection outbreaks in our hospitals. Ergonomics can offer insights into the causes and consequences of major incidents, from accidents like the Chernobyl

Nuclear Power Plant catastrophe (e.g., Munipov, 1992), the 1999 Ladbroke Grove rail crash (e.g., Stanton and Baber, 2008; Stanton and Walker, 2011) and the Kegworth air disaster (e.g., Griffin et al., 2010; Plant and Stanton, 2012), to the planning and implementation of large-scale public events such as the Olympic Games (e.g., Franks et al., 1983) and relief efforts like that which followed the 2010 Haiti earthquake (e.g., Hutchins, 2011). It can be applied to investigate a hugely diverse, and sometimes surprising, range of issues, from the most optimal search strategies of football goalkeepers facing a penalty shoot-out (Savelsbergh et al., 2005) to the effects of expertise on the performance of crime scene investigators (Baber and Butler, 2012). A recent paper promoting a strategy for the discipline, described Ergonomics as having great potential to optimise the performance and well-being of humans interacting with any designed artefacts, from single products to entire systems and environments (Dul et al., 2012). The authors described developments in the external world that are changing the way humans interact with their environment; these included the introduction of technology with capabilities that far exceed those of the human, significant changes in the way humans interact with technology, and advances in the type of information transferred via new telecommunications and media (Dul et al., 2012). These changes will continue to have a huge impact on the way humans experience the world, including work, travel, healthcare, education, entertainment, and other activities.

The way in which information is transferred to human users via the products they use and the systems they are part of is the focus of HCI research, which could be considered a subdiscipline of Ergonomics. In fact, there has always been a close link between Ergonomics and HCI, for example, in 1982 the first Conference on Human Factors in Computing Systems (CHI) was cosponsored by the Human Factors Society (Grudin, 2009). The emergence of HCI as a discipline coincided with the shift from the use of computers purely in the workplace to the rise of personal computing in the early 1980s (Carroll, 2010; Dix, 2010; Noyes, 2009). The introduction of HCI is attributed to Brian Shackel (see Shackel, 1959) in the late 1950s (Dix, 2010; Dillon, 2009), although it was fully established later with the founding of HCI conferences, including INTERACT and CHI, in the early 1980s (Dix, 2010). Since its conception, HCI has expanded rapidly, commensurate with the vast developments in interactive technologies, the rapid growth in the number of people using computers (Hartson, 1998) and the accompanying expansion and diversification of the concept of usability (Carroll, 2010). Although when it was first introduced, HCI only referred to desktop computing, today the term encompasses a much wider field of study that centres on the relationships and activities between humans and the diverse range of devices with which they interact (Dix, 2010; Carrol, 2010). In this book HCI is used to refer to the computers inside vehicles, which control a wide spectrum of functions ranging from entertainment to climate control, and the emergent behaviour of drivers using these 'computers'.

Usability is a central concept in HCI, and much of the early work on usability evaluation was borne of the frustrations on the part of HCI researchers that usability issues were only ever considered toward the end of the product development process (Lindgaard, 2009). Usability represents an area in which HCI professionals could have a large influence on the design of new technologies (Dillon, 2009); however, for this to be successful, design for usability needs to be supported by effective methods and tools. Dix (2010) differentiated between the goals of usability (i.e., practice, leading to an improved product) and the goals of HCI research (i.e., scientific theory, leading to new understanding), although he also stated that these goals should be interlinked because 'effective design will be based on a thorough understanding of the context and technology' (Dix, 2010; p. 15). The integration of science and practice has been described as a 'significant achievement' of HCI (Carroll, 2010) and is also a key feature of the Ergonomics discipline (Buckle, 2011), although Buckle differed slightly in his description of it as an 'ongoing tension between current practice and … research' (p. 2), which relies on the effective exchange of information between researchers and practitioners. Other authors have echoed Buckle's concerns; for example, Caple (2010) suggested that the beneficiaries of much of the Ergonomics research do not actually read the academic journals or attend the conferences in which it is published. Consequently, it appears that although the combination of both science and practice is seen as very positive for a well-balanced discipline (Buckle, 2011), there is a real need for the correct balance to be struck and to encourage and support effective communication between the researchers generating the knowledge, the practitioners applying this knowledge and also, perhaps most important, the users interacting with the end products. In this vein, Meister (1992) recommended that although Ergonomics is dependent upon research, this research 'must be geared to the … needs of the design engineer'. Caple (2010) summarised these requirements by proposing a holistic approach, engaging a wide range of stakeholders, in order to sustain the effectiveness of Ergonomics interventions. The integration of usability goals with the goals of HCI research, realised by an interlinking of theory and practice, is also a central theme pursued in this book. The knowledge generated via review, analysis and experiment is presented in an accessible way for automotive manufacturers and designers to use to support practical product development.

Long (2010) stressed the importance of a consensus of a particular design problem before successful evaluation can take place. Without this, he suggested, researchers are unable to validate or compare their findings. The concept of usability is an example of where a detailed definition is required before evaluation can take place, although this is a concept which is not straightforward to describe, despite numerous attempts by a number of authors (see Bevan, 1991; Nielsen, 1993; Norman, 1983; Shackel, 1986; Shneiderman, 1992). Although usability is widely regarded to be 'a crucial driving force for user interface design and evaluation' (Hartson, 1998), it is difficult to create a universal definition because it is so dependent on the specific aspects of the context within which particular devices and products are used. In this book, significant attempts at defining usability, along with the issue of context, which has made this such a difficult task, are discussed. As advocated by Long (2010), a conceptualisation for a usable IVIS is presented in the early chapters of this book in the form of usability criteria. This work forms the foundation for a comprehensive evaluation process targeted at understanding and improving the usability of IVIS.

USABILITY EVALUATION

The concept of usability is constantly evolving, and Carroll (2010) suggested that it will continue to do so 'as our ability to reach further toward it improves' (p. 11).

The increasing richness of the concept also means that the evaluation of usability is becoming ever more complex and problematic (Carroll, 2010). HCI has been described as a 'meta-discipline' (Carroll, 2010), which has always drawn from other fields, including Ergonomics, cognitive psychology, behavioural psychology, psychometrics, systems engineering, and computer science (Hartson, 1998), and this list continues to diversify as usability subsumes wider qualities such as aesthetics, well-being, quality of life and creativity (Carroll, 2010; Hancock and Drury, 2011; Lindgaard, 2009). The context within which products are being used also continues to evolve (Lindgaard, 2009); therefore, researchers need better means of predicting these situations and the associated emergent behaviours. Hartson (1998) reported a general agreement in the literature that interaction development needs to involve usability evaluation: this reiterates the requirement for an integrated and iterative process of design–evaluation–redesign, which is the model presented in this book.

Dix (2010) described evaluation as 'central' to HCI; however, the real or perceived costs associated with evaluation often prevent it from being successfully and sufficiently implemented in the product development process (Hartson, 1998). In reality, usability engineering often does not increase product development costs as much as people may think and the benefits of good usability in the final product will almost always outweigh the costs of the process (Hartson, 1998). A key aim for enforcing this message to manufacturers is to present them with appropriate tools to support usability evaluation, particularly at stages in the process when maximum benefits to the final product will be realised, i.e., early in concept development.

One of the most significant developments in the evaluation of HCI has been the modelling work of Card et al. (1983), which began with the Model Human Processor (MHP) and the Goals, Operators, Methods, and Selection Rules (GOMS) model, and was one of the first applications of cognitive theory in HCI (Carroll, 2010; Hartson, 1998). This work has since been added to with significant contributions in the form of EPIC (Executive Process Interactive Control; Kieras and Meyer, 1997), ACT-R (Atomic Components of Thought; Anderson and Lebiere, 1998) and CPA (Critical Path Analysis; Lockyer, 1984). Rather than measuring human–computer interaction as it occurs, modelling predicts the success of the interaction based purely on prior information about human processing, the tasks that would be performed, and the environment in which they would be performed. Meister (1992) stressed the importance in Ergonomics of being able to *predict* human performance, and it provides a solution to the problems commonly associated with usability testing because it is relative inexpensive to carry out and can be applied at an early stage in the product development process. In this book, a modelling approach is investigated as part of a toolkit of measures for evaluating the usability of IVIS.

BOOK OUTLINE

The book is organised in nine chapters, starting with an introduction which describes the background to the work and outlines the main research objectives (Chapter 1). An overview of the remaining chapters follow.

Chapter 2: Context-of-Use as a Factor in Determining the Usability of In-Vehicle Information Systems. In recent years, the issue of usability of IVIS has received growing attention. This is commensurate with the increase in functionality of these devices, which has been accompanied by the introduction of various new interfaces to facilitate the user–device interaction. The complexity and diversity of the driving task presents a unique challenge in defining usability: user interaction with IVIS creates a 'dual task' scenario, in which conflicts can arise between primary and secondary driving tasks. This, and the safety-critical nature of driving, must be accounted for in defining and evaluating the usability of IVIS. It is evident that defining usability depends on the context-of-use of the device in question. The aim of the work presented in Chapter 2 was therefore to define usability for IVIS by selecting a set of criteria to describe the various factors that contribute to usability in this specific context-of-use and to define Key Performance Indicators (KPIs) against which usability could be measured.

Chapter 3: In-Vehicle Information Systems to Meet the Needs of Drivers. IVIS integrate most of the secondary functions available within vehicles. These secondary functions are aimed at enhancing the driving experience. To successfully design and evaluate the performance of these systems, a thorough understanding of the task, user, and system, and their interactions within a particular context-of-use, is required. Chapter 3 presents a review of these three variables in the context of IVIS, which aims to enhance understanding of the factors that affect system performance. An iterative process for modelling system performance for the task–user–system interaction is also illustrated. This will support designers and evaluators of IVIS in making predictions about system performance and designing systems that meet a set of criteria for usable IVIS.

Chapter 4: A Usability Evaluation Framework for In-Vehicle Information Systems. Usability must be defined specifically for the context-of-use of the particular system under investigation. This specific context-of-use should also be used to guide the definition of specific usability criteria and the selection of appropriate evaluation methods. There are four principles that can guide the selection of evaluation methods, relating to the information required in the evaluation, the stage at which to apply methods, the resources required, and the people involved, for instance, the skills of the analysts and whether or not representative users are tested. Chapter 4 presents a flowchart to guide the selection of appropriate methods for the evaluation of usability in the context of IVIS. This flowchart was used to identify a set of analytic and empirical methods which are suitable for IVIS evaluation. Each of these methods has been described in terms of the four method selection principles.

Chapter 5: The Trade-Off between Context and Objectivity in an Analytic Evaluation of In-Vehicle Interfaces. Chapter 5 presents a case study to explore an analytic approach to the evaluation of IVIS usability, aimed at an early stage in product development with low demand on resources. Five methods were selected: Hierarchical Task Analysis (HTA), Multimodal Critical Path Analysis (CPA), Systematic Human Error Reduction and Prediction Approach (SHERPA), Heuristic Analysis, and Layout Analysis. The methods were applied in an evaluation of two IVIS interfaces: a touch screen and a remote controller. The findings showed that there was a trade-off between the objectivity of a method and consideration of the context of use; this has implications for the usefulness of analytic evaluation. An extension to the CPA method is proposed as a solution to enable more objective comparisons of IVIS, whilst accounting for context in terms of the dual-task driving environment.

Chapter 6: To Twist or Poke? A Method for Identifying Usability Issues with Direct and Indirect Input Devices for Control of In-Vehicle Information Systems. IVIS can be controlled by the user via direct or indirect input devices. In order to develop the next generation of usable IVIS, designers need to be able to evaluate and understand the usability issues associated with these two input types. The aim of the study presented in Chapter 6 was to investigate the effectiveness of a set of empirical usability evaluation methods for identifying important usability issues and distinguishing between the IVIS input devices. A number of usability issues were identified and their causal factors have been explored. These were related to the input type, the structure of the menu/tasks, and hardware issues. In particular, the translation between inputs and on-screen actions and a lack of visual feedback for menu navigation resulted in lower levels of usability for the indirect device. This information will be useful in informing the design of new IVIS, with improved usability.

Chapter 7: Modelling the Hare and the Tortoise: Predicting IVIS Task Times for Fast, Middle and Slow Person Performance using Multimodal Critical Path Analysis. Analytic models can enable predictions about important aspects of the usability of IVIS to be made at an early stage of the product development process. Task times provide a quantitative measure of user performance and are therefore important in the evaluation of IVIS usability. In this study CPA was used to model IVIS task times in a stationary vehicle and the technique was extended to produce predictions for 'slowperson' and 'fastperson' performance, as well as average user ('middleperson') performance. The CPA-predicted task times were compared to task times recorded in an empirical simulator study of IVIS interaction and the predicted times were, on average, within acceptable precision limits. This work forms the foundation for extension of the CPA model to predict IVIS task times in a moving vehicle, to reflect the demands of the dual-task driving scenario.

Chapter 8: Visual Attention on the Move: There Is More to Modelling Than Meets the Eye. The use of CPA to predict single-task IVIS interaction times was demonstrated in Chapter 7. The aim of the study presented in Chapter 8 was to investigate how the CPA model could be extended for accurate prediction of dual-task IVIS interaction times, for instance, tasks performed at the same time as the primary driving task. Two models of visual behaviour were proposed and tested against empirical IVIS task times: one model tested the 'separate glances' hypothesis whilst the other tested the 'shared glances' hypothesis. The model that incorporated 'shared glances', in which visual attention is used to obtain information from both the IVIS and road scene simultaneously, produced the most precise predictions of IVIS task time. The findings of this study raise important questions about the division of visual attention between primary and secondary tasks. It appears that peripheral visual monitoring can be utilised in a dual-task environment, although it is likely that certain types of visual information are more suited to peripheral processing than others. Further investigation of shared glances will improve the precision of future dual-task HCI models and will be useful in the design of interfaces to enable peripheral processing.

Chapter 9: Summary of Contributions and Future Challenges. Chapter 9 summarises the work presented in this book and explores the findings using a number of key questions that arose during the project. The implications of the research are discussed along with the future challenges in the area of IVIS design and evaluation.

The work presented in this book contributes to the understanding and evaluation of usability in the context of IVIS. The definitions and criteria for usability in an IVIS context will be useful to structure future studies of driver-vehicle interactions and in the development of new interaction strategies. The toolkit of analytic and empirical evaluation techniques was based on a comprehensive review of Ergonomics methods: this will provide a valuable reference tool, offering information not only on the output of various methods, but also on their utility at various stages throughout the product design process. The IVIS evaluation case studies have identified usability issues which limit the success of current interaction strategies and have highlighted the importance of optimisation between individual components of HMI. The CPA method was extended for quantitative predictions of IVIS interaction times in both stationary and moving vehicle situations: this was targeted at automotive manufacturers to address a need for early-stage product evaluation. The 'shared glance' hypothesis, which was developed as a result of work on the CPA model, contributes to the knowledge of visual processing in dual-task environments. Modelling the visual aspect of the driver-IVIS interaction more precisely will result in more accurate predictions of the effect of IVIS use on driving. This information will be useful in the development of more usable IVIS, with the goal of enhancing the driving experience and reducing distraction.

2 Context-of-Use as a Factor in Determining the Usability of In-Vehicle Information Systems

INTRODUCTION

The first references to the concept that is now most commonly known as 'usability' used terms, such as 'ease of use' (Miller, 1971, cited in Shackel, 1986), 'user friendliness' (Dehnig et al., 1981), and 'user-perceived quality' (Dzida et al., 1978). It was widely thought that these terms created a narrow view of the concept in which the person is treated as a single system component (Adler and Winograd, 1992; Bevan, 1991). This traditional view was criticised for overlooking users' cognitive and social characteristics and not considering the processes of learning and adaptation to systems and products (Adler and Winograd, 1992). It also suggested that usability is a characteristic that can simply be designed into a product (Bevan, 1995) and failed to account for other influencing factors, such as users' past experiences and their expectations and attitude, as well as the features of the product itself (Baber, 2002). In response to this criticism and to calls for a more precise definition (Norman, 1983), the term *usability* was adopted, with the first attempt at a definition being widely attributed to Brian Shackel in 1981 (Baber, 2002; Shackel, 1986). Early definitions of usability were based on the usability of computer software (Dehnig et al., 1981; Long, 1986; Ravden and Johnson, 1989; Sweeney et al., 1993). This is because the term was most commonly associated with the field of HCI. In the 1970s and 1980s, people only encountered computers at work, and so definitions of usability dealt primarily with work contexts. More recently, however, the gulf between computers and 'ordinary people' has reduced dramatically (Cox and Walker, 1993), and definitions of usability have widened to encompass any product or system with which a user interacts, whether for work or for leisure purposes. Bevan (1999) documented this transition in the view of usability from computer-related to a broader view, and added a final stage to the development of usability: the realisation that usability should be a central goal of design. This reflects the increase in the importance of this concept since it was first defined, which has been driven by a decline in users' acceptance of poor design and the increasing complexity of products (Stanton and Young, 2003).

SIGNIFICANT CONTRIBUTIONS TO DEFINING USABILITY

There have been a number of significant contributions to the definition of usability, and these are summarised in Figure 2.1. The first formal definition of usability was attributed to Brian Shackel whose work introduced usability as a quantifiable concept. Donald Norman focussed more on the user's perspective, in particular, on the 'conceptual model' a person creates of a particular product and how this must be considered in design for usability. Like Shackel and Norman, Jakob Nielsen presented a quantitative approach to usability, introducing 'usability engineering' as a systematic method for the evaluation of usability. Ben Shneiderman also focussed more on the evaluation side of usability in the specification of his 'eight golden rules of dialog design' and five 'human factors goals', which were also firmly rooted in the existing usability theory. Nigel Bevan was responsible for collating much of the work on usability and collaborating with the ISO to develop standards relating to usability.

BRIAN SHACKEL

Shackel (1986) stated that usability can be defined by the interaction between the user, the task, and the environment. Shackel was the first to emphasise the need for a 'definable specification' (Shackel, 1997), and his work was important in identifying usability as a 'key concept'. He defined four factors to describe a usable system: effectiveness, learnability, flexibility, and attitude. Shackel's 'formal operationalised definition of usability' was particularly important because it was the first to propose the application of quantitative techniques to the evaluation of usability. He proposed that numerical values should be assigned to various attributes of each of the four

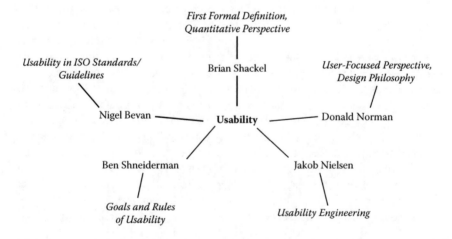

FIGURE 2.1 Significant authors and their contributions to defining usability.

usability factors. Defining these attributes allow designers to specify exactly the level of performance required to achieve usability. Shackel's approach to defining usability has been criticised by some for being too restrictive (Cox and Walker, 1993) because of its focus on quantitative attributes of usability. The ambiguous nature of specifying values for levels of performance on tasks and the likely low consistency of these values between different tasks and different users are also potential problems with this definition. This was the first time that the definition and evaluation of usability had been formally addressed however, and it paved the way for much more work in the area. Since they were first proposed by Shackel, the four factors of usability have been incorporated into many further definitions. Stanton and Baber (1992) re-described the factors under the acronym LEAF; however, they criticised this definition for excluding factors that they considered 'equally important'. To account for these deficiencies they added four extra factors of usability, based on work by Eason (1984) and Booth (1989). Probably the most significant of these additional factors was 'the perceived usefulness or utility of the system', which was inspired by Booth's (1989) comment that a system which is rated highly according to LEAF may not necessarily be used in real life. The importance of this criterion has since been acknowledged in other definitions and discussions of usability (Cox and Walker, 1993; International Organization for Standardization, 2006). The three further criteria defined by Stanton and Baber (1992) were task match, task characteristics, and user criteria. This revised definition goes further to address Shackel's original statement that usability is defined by the task, user, and environment.

Jakob Nielsen

Nielsen defined usability as 'a quality attribute that assesses how easy user interfaces are to use' (2009). He described five components of usability: learnability, efficiency, memorability, errors, and satisfaction (Nielsen, 1993). Nielsen defined these factors as precise and measurable components of the 'abstract concept of usability', arguing that a systematic approach to usability was required, and criticising the term 'user friendly' for being inappropriate and too narrow. He referred to this systematic approach as 'usability engineering'. Although Nielsen and Shackel only listed one common factor (learnability) in their definitions of usability, there is much overlap between the two descriptions. For example, memorability (Nielsen) is related to learnability (Shackel and Nielsen); efficiency (Nielsen) is a measure of effectiveness (Shackel) against some other metric such as time; errors (Nielsen) are closely linked to effectiveness and efficiency (Shackel); and satisfaction (Nielsen) is synonymous with attitude (Shackel). This is evidence of the difficulty in defining concrete terms for usability and is perhaps one reason why a universal definition of usability has so far proved difficult.

Nielsen also classified usability, alongside utility, as an attribute of usefulness, which itself was an attribute of practical acceptability (Nielsen, 1993). In this way he distinguished between utility and usability, describing the former as an issue of the functionality of a product in principle and whether this allows the product to perform in the way that it is required. He described usability in relation to this as 'how well users can use that functionality' (1993, p. 25).

This decision to treat usability and utility as separate is typical of many definitions and is a significant step in defining usability as a distinct concept, because it demonstrates the shift in focus from product-centred design, which relates to the functionality of a product (that is utility to user-centered design how well the user is able to use that functionality). Nielsen (1993) proposed the possibility of an analytic method that could be used to design usable products based on a set of usability goals. He also suggested that there may exist some interaction techniques that would solve the problem of usability because they are so inherently easy to use, citing speech input as a possible example. This prediction was somewhat naïve, given the problems with speech input technologies and the highly variable levels of user satisfaction associated with this type of interaction. It is therefore unlikely that a product interface for a usable product will ever suit all situations, tasks, and users.

DONALD NORMAN

Like Shackel, Norman also acknowledged the need for a more precise definition of usability in his earlier work in the area, stating that it was not enough to instruct designers to just 'consider the user' (Norman,1983). Further evidence of this willingness to move away from vague concepts such as ease of use and user friendliness can also be found in a book by Norman and Draper (1986), which under the index term 'user friendliness', has the line 'this term has been banished from this book'. One of the major focusses of Norman's work in usability was on the user's perspective, which was important in shifting the emphasis of the more traditional definitions from the product to the user.

Later, with the shift in focus from product to user, Norman (2002) proposed his 'principles of design for understandability and usability', which consisted of advice to provide 'a good conceptual model' and to 'make things visible'. Norman suggested that forming an accurate conceptual model of an interaction would allow users to predict the effects of their actions. He made the connection between the user's conceptual model and the designer's conceptual model of a product, suggesting that the system image, that is, the parts visible to the user, must consolidate the designer's model of the interaction with the user's expectations and interpretations of the interaction. According to Norman (2002), this system visibility consisted of two components: mapping and feedback. Mapping refers to the relationship between the controls and the effects of interacting with them, and feedback is the principle of sending back information to the user about the actions that have been performed. He proposed that if these two components are capable of portraying an accurate and adequate representation of the product to the user, then the product will have high understandability and usability. Norman (2002) suggested that this could be achieved by balancing the 'knowledge in the world' with the 'knowledge in the head'. He recommended that the required knowledge should be put in the world, that is, by ensuring good system visibility. Knowledge in the head can then be used to enhance interactions once the user has learned the relevant operations. Norman collected these principles of design into a 'design philosophy' which focussed on 'the needs and interests of the user' and therefore supported a more user-centred design process. Norman (2002) proposed some further design principles in the form of his instructions on 'how to do things wrong'. This was a list of what not to do when designing for usability and included 'make things invisible', 'be inconsistent' and 'make operations unintelligible' (Norman, 2002). These were based on his original 'principles of design for understandability and usability', although they were more instructional in nature. The benefit of this approach was that listing what not to do highlighted the mistakes that could be made if the user and usability were not considered in the design process. Norman (2002) also defined 'seven principles for transforming difficult tasks into simple ones'. As well as including further principles referring to visibility and mapping, these principles recommended designing for error (i.e., assuming that any mistake that could be made will be made), simplifying the structure of tasks and using standards when necessary. There is a great deal of overlap between Norman's various sets of design principles, and this makes interpreting his work fairly difficult. His 'seven principles for transforming difficult tasks into simple ones' are probably the most suitable to guide designers because they incorporate all of the factors that Norman identified as important in usability. The output of Norman's work on usability is perhaps better viewed as a contribution to the philosophy of usability, rather than as a definitive list of usability criteria.

BEN SHNEIDERMAN

Shneiderman (1992) agreed with criticisms that earlier terms, such as *user friendliness*, were too vague and suggested that designers needed to go beyond this to produce successful products. He also discussed the importance of context and suggested that this affects the importance of various attributes of usability. For example, he proposed that for office, home, and entertainment applications, the most important usability attributes would be ease of learning, low error rates, and subjective satisfaction. This is in contrast to commercial use of products and systems, in which user satisfaction may be less important because use is not voluntary. Shneiderman (2000) also discussed the issue of user diversity, highlighting the need to understand differences between users in terms of skill, knowledge, age, gender, ability, literacy, culture, and income. He described this 'universal usability' as important, given the high level of interactivity that was a product of the growth in HCI applications for information exploration, commercial applications, and creative activities (Shneiderman, 2000).

In terms of a definition of usability, the closest Shneiderman (1992) offered was his '8 golden rules of dialog design'. This list included 'Strive for consistency', 'Offer informative feedback' and 'Permit easy reversal of actions'. Rather than criteria for usability, these rules are more similar to design guidelines. Shneiderman (1992) also defined five 'human factors goals' against which usability could be measured. Although these goals describe the measurable aspects of usability, and are not a direct definition, there is much overlap with Nielsen's five attributes of usability and Shackel's LEAF precepts.

NIGEL BEVAN AND THE INTERNATIONAL ORGANIZATION FOR STANDARDIZATION (ISO)

Nigel Bevan contributed to the development of ISO 13407, Human-Centred Design Processes for Interactive Systems (International Organization for Standardization,

1999), which provides guidance on human-centred design activities, including designing for usability. He has also written extensively on other usability-related ISO standards, and his work is therefore discussed here alongside that of the ISO. Today, the most commonly cited definition of usability is probably that found in ISO 9241, Ergonomic Requirements for Office Work with Visual Display Terminals (VDTs)— Part 11: Guidance on Usability (International Organization for Standardization, 1998). In this standard, usability is defined as

> [The] extent to which a product can be used by specified users to achieve specified goals with effectiveness, efficiency and satisfaction in a specified context of use. (1998, p. 2)

The reason for the wide adoption of this definition is probably the inclusion of the term *context-of-use*. The standard places much emphasis on this, stating that usability is dependent on context, that is, 'the specific circumstances in which a product is used'. Consideration of the context-of-use makes a general definition of usability virtually impossible because different situations will demand different attributes from a product to optimise the interaction with particular users. Despite the desire to construct a universal definition, it appears that most people now accept that the context in which a product or system is used must be taken into account, and definitions, therefore, need to be constructed individually according to the product, tasks, users, and environment in question. Heaton (1992) suggested that within the context of a particular product, an explicit definition of usability can be developed and used for evaluation of the product.

The inclusion of efficiency in the ISO definition is particularly useful for the evaluation of usability because it relates effectiveness, that is, how well the user is able to accomplish tasks, to the expenditure of resources, such as human effort, cost, and time, and therefore can be measured relatively easily. User satisfaction is less easy to interpret as it is linked to user opinion and can therefore only be assessed subjectively. Subjective satisfaction has, however, often been incorporated into definitions of usability (i.e., Shackel, Nielsen, Norman), and this reflects its importance as an aspect of usability. In an extension of the ISO definition, Kurosu (2007) distinguished between the subjective and objective properties of usability, and suggested that satisfaction is the user's subjective impression of the other two ISO criteria of usability: effectiveness and efficiency. It is not only the perception of effectiveness and efficiency that is important, however; there are many more criteria that contribute to subjective satisfaction, including aesthetic and emotional appeal. It is likely that for a comprehensive assessment of subjective satisfaction, many aspects would need to be evaluated; however, these aspects are not defined in ISO 9241, a problem which has led to criticisms that it is too broad (Cacciabue and Martinetto, 2006; Jokela et al., 2003). In particular, Baber (2002) considered the exclusion of factors such as pleasure, fun, and coolness to be a weakness of this definition. The omission of learnability has also been a cause of criticism of ISO 9241 (Noel et al., 2005), particularly because this aspect of usability has been considered so important by others, including Shackel (1986) and Nielsen (1993), who described it as 'the most fundamental usability attribute'. Butler (1996) also identified learnability as 'a critical aspect of usability', reasoning that learning how to use a system is the first, and therefore one of the most important, interactions a user has with it.

An alternate ISO standard was developed separately from ISO 9241, and is aimed specifically at software product quality. ISO 9126, Information Technology—Software Product Quality—Part 1: Quality Model (International Organization for Standardization, 2001) refers to six attributes of 'external and internal quality' of a software product, one of which is usability. ISO 9126 defined usability as

> The capability of the software product to be understood, learned, used and attractive to the user, when used under specified conditions. (2001, p. 9)

Like the ISO 9241 definition, this also refers to context-of-use as a factor that determines usability of a certain product or system. However, there are distinct differences between the usability factors described in each definition; for example, ISO 9126 defines understandability, learnability, and attractiveness as factors contributing to usability, whereas ISO 9241 lists efficiency, effectiveness, and satisfaction. The term *used* in the ISO 9126 definition is also ambiguous as it seems to suggest that capability of 'use' is a sub attribute of usability; however, these terms are not clearly distinct. On the other hand, 'use' may also be considered synonymous with 'utility', which is usually considered a distinct concept to usability, rather than an attribute of it. A second concept used in ISO 9126 is 'quality in use', which is defined as

> The capability of the software product to enable specified users to achieve specified goals with effectiveness, productivity, safety, and satisfaction in a specified context of use. (2001, p. 12)

This is more similar to the definition of usability in ISO 9241, as it refers to effectiveness and satisfaction, although it also introduces productivity and safety as additional factors. When examined as a whole, there is some overlap between the definitions presented in ISO 9126 and 9241, and Bevan (2001) recommended that they be seen as complementary. Bevan (2001) attributed the differences between the two definitions of usability to the fact that ISO 9241 takes a much broader view of the concept. He suggested that the two standards need to be combined, and this could be useful in the development of a standard for the usability of all types of interactive systems and products, to address the deficiencies in the scope of application of current standards.

A final ISO standard applied the definition of usability from ISO 9241 to the use of 'everyday products'. In ISO 20282, Ease of Operation of Everyday Products—Part 1: Design Requirements for Context-of-Use and User Characteristics (International Organization for Standardization, 2006), effectiveness is said to be the most important attribute of usability when applied to everyday products. This is because interaction with these products is 'generally fast and of low complexity'. This standard refers to 'ease of operation', rather than usability – and defines the former as

> usability of the user interface of an everyday product when used by the intended users to achieve the main goal(s) supported by the product. (2006, p. 2)

This definition just adds to the confusion surrounding the issue of usability as defined in standards and perhaps re-enforces Bevan's (2001) call for a standard that defines usability and related concepts for all types of interactive products and systems, although whether or not this is a feasible goal is questionable. This will be discussed later in the chapter.

THE DEVELOPMENT OF A DEFINITION OF USABILITY

The work of the five authors discussed above has supported the evolution of an approach to evaluating usability which has developed over the last 30–40 years. Bevan (1991) attempted to categorise this evolution of usability according to four classifications: product-oriented, user-oriented, user-performance view, and contextual view. Very early definitions of usability had a distinct product-focus, or 'engineer's view' (Stanton and Baber, 1992) and implied that usability could simply be designed into a product. Shackel's early, quantitative view of usability reflects a product focus, in which only measurable attributes of a product were considered to be of importance. Although Nielsen's definition of usability shared many qualities with Shackel's, Nielsen (1993) also acknowledged the importance of users' individual characteristics and differences in understanding usability. This reflects the realisation that the user was central to design and was accompanied by the appearance of more user-oriented definitions, which tended to focus on the workload imposed by a system on the user. For example, Norman advocated a focus on the user's conceptual model of a product and task, and more subjective aspects of usability, reinforcing the importance of considering the user in design. Finally, as usability was developing as a concept, it became obvious that the context-of-use was of utmost importance (see definitions by Bevan, 1991; International Organization for Standardization, 1998; Shneiderman, 1992). Chamorro-Koc et al. defined context-of-use as 'the relationship between the use-activity-situation during people's interaction with products' (2008, p. 648).

Context-of-use defines not only the attributes of the product that are important in determining usability, but also the tasks being performed using the product, the users performing those tasks, and the environment within which the tasks are being performed.

A Universal Definition of Usability?

Gray and Salzman (1998) likened efforts at creating a clear definition of usability to 'attempts to nail a blob of Jell-O to the wall'. The evidence presented here shows that there is unlikely ever to be a single universally accepted definition of usability because the issue of context is so important. Most definitions of usability include some reference to the context-of-use of a product or system, and most attributes of usability will vary in importance depending on the context. For example, Nielsen (1993) and Noel et al. (2005) proposed that memorability was important for usability, however the level of importance will be dependent on the context-of-use. Memorability will be more significant for products and systems that are used infrequently compared to those that are used on a daily basis, because high frequency of use improves information retention. The issue of context is perhaps most clearly described by Bevan (2001) who stated that a product does not have any 'intrinsic usability'; rather, it has 'a capability to be used in a particular context'. Stanton and Baber (1992) suggested that the development context of the product should also be considered in defining usability because a designer's view of usability as defined at the concept stage of design may well be vastly different to a user's view of the end product. In defining usability for a product the stage of development must therefore be taken into account and appropriate criteria must be selected.

Despite the interest in defining usability, many papers on the subject have neglected to provide any definition or explanation, even when it is referred to throughout the document (e.g., Barón and Green, 2006; Gould and Lewis, 1985). It seems that most authors acknowledge the need for usability and good design; however, some do not specify what this actually means (Dehnig et al., 1981). This is because most people have an understanding of what usability is but have difficulty in defining it in a useful way (Stanton, 1998). This only compounds the problem of defining usability because it is open to misinterpretation if not fully described in the context of the work. Lansdale and Ormerod (1994) also suggested that one of the difficulties with defining and evaluating usability is the fact that it is easier to identify when it is absent than when it is present; they therefore described usability as an 'anti-concept'. This is not to say that usability is a 'hygiene' factor (Herzberg, 1996); that is, its absence leads to dissatisfaction but its presence does not lead to satisfaction. Rather, better definition of the concept would lead to easier identification of its presence.

There have also been many further additions to the main usability criteria identified previously in this review and this has contributed to the problem of defining the concept in a single set of attributes. Baber (2005b) identified 34 factors of usability, highlighting the difficulty in defining such a complex concept. A number of authors have suggested that aesthetics should be considered a component of usability (Chestnut et al., 2005; Lindgaard and Whitfield, 2004; Macdonald, 1998; Preece et al., 2002), although the importance of this is a matter of some debate and Lindgaard and Whitfield (2004) reported that human factors papers are 'virtually devoid' of references to aesthetics. Khalid and Helander described aesthetics as 'an attractive look, touch, feel and attention to detail' (2004, p. 30). It could be argued that a product with low aesthetic value is still usable; however, it is still a very important factor in the appeal of a product and therefore in determining its perceived usefulness (Stanton and Baber, 1992).

Additional usability criteria also include naturalness, advanced feature usage (Hix and Hartson, 1993); helpfulness, motivation, emotional fulfilment, support for creativity, fun to use (Preece et al., 2002); intuitiveness, supportiveness, controllability, avoidance of physical and mental load (Maguire et al., 1998), and replaceability, portability, and recoverability (Baber, 2005b). Many of these additional attributes are subjective in nature. Subjective criteria have been less well defined in the main definitions of usability, perhaps because of the difficulty in measuring such attributes. This does not mean, however, that they are any less important to the usability of a product or system, and the inclusion of these attributes in more recent definitions reflects that this is beginning to be realised. Khalid and Helander (2004) warned that subjective evaluation may be affected by the experience a user has of the product being tested.

They hypothesised that a novice user may focus only on the 'holistic impression and styling', whereas a user with more experience will be able to offer opinions on a wider range of aspects. The literature contains hundreds of suggestions of factors that constitute usability, and each one is, to some extent, correct. However, these factors will not all be applicable or appropriate in every circumstance and will vary in importance depending on the context. The challenge is to decide which factors are appropriate in each case to ensure that all products have a high level of usability.

SPECIFYING USABILITY FACTORS

The key factors which are integral to any definition of usability, as proposed by the five authors identified previously, are presented in Table 2.1. The extension of Shackel's LEAF precepts, proposed by Stanton and Baber (1992), are also included in the table as these criteria are not covered by the other authors' definitions and they are considered to be equally relevant.

These lists of factors were synthesised to produce a single list of high level usability factors. The three factors specified in ISO 9241-11 (effectiveness, efficiency, and satisfaction) are widely used, and they were selected first. Shackel's 'attitude' factor was considered to be synonymous with 'satisfaction' and was not included in the list. Additional factors of learnability (Shackel and Nielsen), flexibility (Shackel), memorability (Nielsen), and Stanton and Baber's additional factors, were also included in the list. Error (Nielsen, Shneiderman) was considered to be a contributing criteria to efficiency and therefore wasn't included in the high-level factor list. Norman's seven principles were considered to represent a design philosophy rather than individual usability factors, and therefore were not included in the list. The final list of high-level usability factors is as follows:

- Effectiveness
- Efficiency
- Satisfaction
- Learnability
- Memorability
- Flexibility
- Perceived usefulness
- Task match
- Task characteristics
- User criteria

These ten high-level usability factors are loosely defined and need to be considered in terms of the context of use of a particular product or system in order to specify more detailed, and therefore more useful, criteria and Key Performance Indicators (KPIs). This book focusses on a case study of IVIS. In order to evaluate these systems, the usability factors need to be related to the context of use of IVIS, that is, driving, so that more specific usability criteria and KPIs can be specified to guide systematic assessment. IVIS and their associated context of use are discussed in the following sections of this chapter.

TABLE 2.1
Key Usability Factors Proposed by the Significant Authors in the Field

Brian Shackel	Jakob Nielsen	Donald Norman	Ben Shneiderman	Nigel Bevan/ ISO
Learnability	Learnability	Use knowledge in the world and in the head	Time to learn	Effectiveness
Effectiveness	Efficiency	Simplify the structure of tasks	Speed of performance	Efficiency
Attitude	Memorability	Make things visible	Rate of errors	Satisfaction
Flexibility	Errors	Get the mappings right	Retention over time	
Stanton and Baber	Satisfaction	Exploit the power of natural and artificial constraints	Subjective satisfaction	
Perceived usefulness		Design for error		
Task match		When all else fails, standardise		
Task characteristics				
User criteria				

USABILITY OF IN-VEHICLE INFORMATION SYSTEMS (IVIS)

Part One of this chapter has identified the high-level factors that are common to general definitions of usability. These factors are not particularly useful for supporting usability evaluation until they are specified in more detail in the form of criteria and KPIs. In order to do this, an understanding of the context of use within which a particular product or system is used is required. The second part of this chapter is therefore concerned with defining this context of use for IVIS.

In this book, the term 'IVIS' refers to a screen-based interface within a vehicle, which incorporates most of the secondary functions available to the driver. Secondary vehicle functions relate to the control of communication, comfort, infotainment, and navigation, whereas primary functions are those involved in maintaining safe control of the vehicle (Lansdown, 2000), that is, the driving task. In recent years, the issue of usability of IVIS has increased in importance. This is commensurate with the increases in the functionality offered by these devices and with the subsequent realisation that this creates a potential source of distraction to drivers, with significant risk to safety.

Defining the Context-of-Use

There have been a number of attempts to define usability within the context of IVIS, in particular, as a foundation for developing guidelines to assist the design and evaluation of such devices (e.g., Alliance of Automobile Manufacturers, 2006; Commission of the European Communities, 2008; Japan Automobile Manufacturers Association, 2004; Stevens et al., 2002; The European Conference of Ministers of Transport,

2003). The preliminary step involved in developing any definition of usability should be to define the context-of-use of the particular product or system involved. The context-of-use within which an IVIS must be defined is perhaps more important than many other products because it is closely linked to additional safety-critical interactions, and the impact on these must be carefully considered. Fastrez and Haué (2008) suggested that the high diversity of the driving context also increases the complexity of designing for usability, compared with other products and systems. A review of the literature was conducted to explore the context-of-use for IVIS. The context-of-use is dependent on a number of factors, including (1) the users involved, that is, the range of user characteristics typical of the driver population that will influence IVIS interaction, the need for training to match the users' capabilities with task demands, and the influence of users' perceptions of the IVIS on product appeal and saleability; (2) the environment in which the interaction takes place, that is, the impact of interacting with an IVIS whilst simultaneously driving on performance of both tasks and the effect of conditions such as night-driving or excessive sunlight on the user-IVIS interaction; and (3) the tasks being performed, that is, the IVIS functions which are used on a regular basis by drivers (Amditis et al., 2006; Fuller, 2005; Hedlund et al., 2006; Lansdown et al., 2002; Stevens et al., 2002; Young et al., 2009a). Six major context-of-use factors were identified from the literature on IVIS, specifically to reflect the importance of considering the users, environment and tasks in an analysis of usability:

- Dual task environment
- Environmental conditions
- Range of users
- Training provision
- Frequency of use
- Uptake

A thematic analysis was conducted in the context of IVIS to identify the main issues which influence usability. These issues were categorised under the six headings listed above. The sources for this analysis consisted of studies of the usability of IVIS and guidelines for their design and evaluation (e.g., Alliance of Automobile Manufacturers, 2006; Commission of the European Communities, 2008; The European Conference of Ministers of Transport, 2003; International Organization for Standardization, 1996; Japan Automobile Manufacturers Association, 2004). The justification for these choices is presented below.

Dual Task Environment

Fastrez and Haué (2008) suggested that one of the most important contextual factors in defining usability for IVIS is the fact that interacting with these devices is not usually the user's main task, that is, the majority of the time they will also be performing the primary task of driving the vehicle (Stanton and Young, 1998a). This distinguishes IVIS from most other products and systems in terms of the context-of-use, and introduces the problem of considering a 'dual task environment' (Burnett, 2000; Lansdown et al., 2002) in designing for usability. Efficiency, which is defined as one of the three main aspects of usability in ISO 9241 (International Organization for Standardization, 1998), is an important measure in this dual task environment as it can indicate conflicts between primary driving performance and interaction with an IVIS. Efficiency in this sense is a measure of effectiveness (another ISO 9241 factor) against the effort expended by the driver, and is important because the amount of effort focussed on using the IVIS is inversely proportional to the amount of effort left for use in performing the driving task. If too much effort and attention is diverted from the primary driving task to the interaction with the IVIS then driving performance will be degraded, resulting in potential risks to safety (Endsley, 1995; Matthews et al., 2001). The effects of interacting with an IVIS on performance of the driving task can also be evaluated in terms of 'interference' (Fastrez and Haué, 2008; International Organization for Standardization, 2003). Interference can be evaluated by comparing the amount of time the user spends performing the primary task of driving against the amount spent interacting with the IVIS. The more time a user spends interacting with the IVIS, the less time they have for performing primary tasks. When the secondary task takes attention away from the primary task the driver is said to be distracted. Driver distraction has become a widely known issue in the past few years as it is linked to reduced safety (Hedlund et al., 2006; Ho and Spence, 2008). The importance of dual task interference and its impact on driver distraction and safety distinguishes IVIS from many other products and systems (Marcus, 2004): this must be of the utmost importance in defining usability for these devices (Landau, 2002). Jordan gives an example to illustrate this:

> whilst lack of usability in a video cassette recorder (VCR) may result in the user recording the wrong television programme, lack of usability in a car stereo may put lives at risk by distracting driver's attention from the road. (1998b, p. 2)

Those definitions which relate usability to safety (e.g., Cox and Walker, 1993; Jordan, 1998b) are therefore most applicable to IVIS, because the risks to the safety of vehicle occupants posed by a device with low usability would be considerable and serious.

Environmental Conditions

Designing an IVIS in the context of the dual task environment helps to address the issues of conflicts within the in-vehicle environment. It is also important to give consideration to the external vehicle environment by ensuring that the device is usable under all environmental conditions (Fuller, 2005; International Organization for Standardization, 1996). This is very relevant in a driving context because vehicles are driven in a wide variety of differing environmental conditions, unlike other products that perhaps may be designed specifically for operation in a relatively stable environment such as a factory or office. For example, an IVIS must be usable at night as well as during daylight, and any visual components must not be adversely affected by glare from sunlight (Alliance of Automobile Manufacturers, 2006; Commission of the European Communities, 2008; Japan Automobile Manufacturers Association, 2004; Stevens et al., 2002). Road and traffic conditions must also be accounted for in the design of IVIS, with focus on the most demanding combinations of conditions (International Organization for Standardization, 1996; Stevens et al., 2002). Usability in this context can be evaluated by measuring the effectiveness of users' interactions with the system under varying conditions and comparing the results across these conditions. Designers should aim for a high level of effectiveness across all conditions with little variation, and the device should be designed to counter any adverse effects resulting from external environmental conditions.

Range of Users

Users of IVIS are drivers and passengers in vehicles that have such a device installed. This is a very large potential user group and consequently will contain a diverse range of physical, intellectual, and perceptual characteristics that need to be recognised in the design and evaluation of these devices (International Organization for Standardization, 1996). The International Organization for Standardization (1996) recommended that the device should be compatible with 'the least able', especially in circumstances like driving, in which it is difficult to define a specific set of user characteristics. Two of the most important user-related factors are age and experience. Older drivers are expected to have some degree of degradation of physiological, sensory, cognitive, and motor abilities (Baldwin, 2002; Herriotts, 2005) and will therefore experience more difficulties in interacting with an IVIS whilst driving. Drivers with little or no driving experience are also likely to be less able to deal with the dual task environment (Stevens et al., 2002) because more of their attention will need to be devoted to performing the primary task correctly. In designing for usability these limitations must be accounted for. Characteristics can also vary within users, that is, dependent on time of day, level of stress or fatigue, etc. These within-user factors also need to be accounted for in design, although they are much more difficult to control for in the evaluation process. Finally, passengers must be considered as potential users of an IVIS. This interaction is not as critical in terms of safety as the driver–device interaction; however, consideration must be given to make the device usable for passengers and also to prevent any conflicts in use between the passenger and driver. Usability of an IVIS by the full range of potential users can be evaluated by assessing device compatibility with these users. This can be achieved by evaluating the effectiveness and efficiency of the interaction across this range of users and ensuring that results are consistently good.

Training Provision

When a person buys a car they are not required to undergo a period of training in order to learn how to successfully operate the IVIS. This would be both impractical and unpopular. Although most IVIS have accompanying instruction manuals that are aimed at training the user, many users do not have the time or inclination to read them before using the device for the first time (Commission of the European Communities, 2008; Llaneras and Singer, 2002), and it would probably be naïve of manufacturers to assume that they do. Landau (2002) also suggested that user acceptance of IVIS is 'extremely good' when little learning is required. In recognition of the low levels of user interest in and acceptance of training manuals, Stevens et al. suggested that designers of IVIS 'should consider the advantages of providing systems where the need for complicated instructions or training is minimal' (2002, p. 14).

This has implications for the design of IVIS because, although most users of in-vehicle systems will begin as novices, they must start using the system well almost immediately (Marcus, 2004). It is therefore important that IVIS have high learnability (Landau, 2002). Learnability is an important factor of usability which can be measured as the time taken to reach an acceptable level of performance. Evaluation of initial effectiveness and efficiency can also give an indication of the usability of a device when first used (Fastrez and Haué, 2008); this must be high in the context of IVIS because of the lack of training provision.

Frequency of Use

The frequency with which an IVIS is used will depend on a number of subfactors, including the exact purpose of the vehicle in question, and the functions which the device is being used to perform. Even a driver who uses their car on a daily basis may rarely interact with many of the functions provided via the IVIS (Burnett, 2000; Commission of the European Communities, 2008). To account for this, memorability for infrequently used functions within the device must be high.

Although it is not the most common context-of-use for IVIS, it is also important to consider the rental car market when designing for usability (Noel et al., 2005). In this context learnability must be a priority criterion of overall usability because the user will want to achieve a high level of usability in a short time and will not have had time to build up any experience of use. Hix and Hartson (1993) and Kurosu (2007) proposed that usability criteria which addressed this temporal aspect of usability, that is, initial and long-term performance, should be included in definitions. Satisfaction must also be high to ensure that an IVIS is used frequently. There is an important distinction here between short- and long-term satisfaction, and this has been identified in some definitions of usability, for example, Hix and Hartson (1993). Short-term satisfaction, that is, how satisfied a user is with a system after initial use, is especially important in ensuring that the user will want to use the device repeatedly. Once the user is using the system habitually, it is important that they experience high levels of long-term satisfaction to ensure that their use of the device remains frequent and prolonged.

Uptake

Use of an IVIS is often not essential to successful control of the vehicle, but is something that a driver can choose to do to enhance the driving experience. In this case, satisfaction will be an important factor because this will influence whether or not a user chooses to use a device repeatedly. A user's experience will only be enhanced if they are satisfied with the interaction. Satisfaction cannot, however, be viewed in isolation, rather trade-offs between various factors of usability (e.g., satisfaction versus operating complexity) must be carefully considered and the design should be 'optimised' (Dehnig et al., 1981). Perceived usefulness is also a significant factor in uptake of IVIS. This is assessed based on users' opinions of the likelihood that they would use the device in reality. Designers must remember that a system may be usable but this is no guarantee that it is useful and will be used by people in the real world. This is an important indicator of the commercial success of a product and will be of particular interest to automotive manufacturers. Perceived usefulness is a difficult attribute to accurately assess because actual usage can only be measured after the device has been released; however, the subjective evaluation of this attribute is useful in predicting the likely behaviour of real users.

DEFINING USABILITY FACTORS FOR IVIS

Defining the context-of-use for a particular product enables designers/evaluators to transform usability factors into more specific usability criteria and KPIs that will support evaluation. Six contextual factors to describe IVIS were defined for this case study: dual task environment, environmental conditions, range of users, training provision, frequency of use, and uptake. Next, the high-level factors that were derived from the general definitions of usability (Bevan, 1991; International Organization for Standardization, 1998; Nielsen, 1993; Norman, 2002; Shneiderman, 1992) were used as guidance to examine each context factor in more detail. This involved investigating how these general usability factors, such as effectiveness, efficiency, and satisfaction (International Organization for Standardization, 1998), applied in an IVIS context. For example, efficiency of the device is important in the dual task environment and in terms of training provision. When viewed within the specific context of the dual task environment, efficiency must be consistently good to ensure that interference between secondary and primary tasks is always low. In the context of training, however, the focus should be on the initial efficiency of the device, i.e., on first use. This is because initial efficiency will indicate how learnable a device is. The 10 high-level usability factors were examined in this way in relation to one or more of the contextual factors. They were then translated into 12 IVIS-specific criteria that cover all aspects of usability in this context-of-use. The original 12 criteria with links to the relevant context factors, through to the 12 translated IVIS-specific criteria are presented in Table 2.1. KPIs are also included for each usability criterion: these describe how the criteria should be measured in terms of IVIS task times, error rates, task structure, input styles, user satisfaction, and driving performance. Methods for measuring these KPIs are described in Chapter 4.

'Effectiveness' and 'efficiency' were linked to the context factors of dual task environment, environmental conditions and training provision because measures of effectiveness and efficiency of the user-IVIS interaction are needed to assess these factors. In the dual task environment, this specifically refers to effectiveness and efficiency during the primary driving task. 'Interference' was included as an additional usability criterion for safety because it addresses the problem associated with the dual task environment, specifically the interaction between secondary and primary tasks. To assess IVIS usability under different environmental conditions, effectiveness and efficiency must be measured under these conditions. Measures of effectiveness and efficiency of the IVIS when used by novice users will also indicate the level of training provision needed in order to make the device usable. 'Learnability' was related to training provision, although it is assessed via measures of effectiveness and efficiency of device interaction with novice users. 'Task characteristics', 'task match', 'user criteria' and 'flexibility' were all linked to the context factor

TABLE 2.2
The Translation from General Usability Criteria to IVIS-Specific Criteria and Links to KPIs

General Usability Factors	Contextual Factors	IVIS Usability Criteria	Key Performance Indicators (KPIs)
Effectiveness	Dual Task Environment	Effectiveness of IVIS whilst driving	1. Task structure and interaction style should be optimal to minimise usability issues whilst driving.
Efficiency		Efficiency of IVIS whilst driving	2. IVIS task times and error rates should be minimised whilst driving.
		Interference between IVIS and driving	3. Interaction with the IVIS should not affect primary driving performance.
Task Characteristics	Environmental Conditions	IVIS effectiveness under varying driving conditions	4. Task structure and interaction style should be optimal to minimise usability issues under all driving conditions.
Learnability		IVIS efficiency under varying driving conditions	5. IVIS task times and error rates should be minimised under all driving conditions.
Task Match	Training Provision	Effectiveness of IVIS with novice users	6. Task structure and interaction style should be optimal to support IVIS interaction for novice users, i.e., the IVIS should be learnable.
User Criteria		Efficiency of IVIS with novice users	7. IVIS task times and error rates should be minimised and usability issues should not be increased for novice users, i.e., the IVIS should be learnable.
Flexibility	Range of Users	IVIS compatibility with full range of users	8. Interaction style and task structure should be designed to support the full range of user characteristics typical of the driver population.
Memorability	Frequency of Use	Short- and long-term satisfaction with IVIS whilst driving	9. User satisfaction on initial use and after prolonged use should be high for all aspects of the interaction.
		Memorability of IVIS interaction	10. IVIS task times and error rates should be minimised even after a period of nonuse of the device.
Satisfaction	Uptake	Satisfaction on first use of IVIS whilst driving	11. User satisfaction on initial use of the IVIS should be high for all aspects of the interaction.
Perceived Usefulness		Perceived usefulness of IVIS in driving	12. Users should report a high likelihood of using the device during real driving.

range of users. These usability factors collectively refer to all the factors of the user and the task, which influence how well they match and were therefore re-described under the IVIS usability criterion 'compatibility'. In this case, the IVIS must be compatible with all potential users of the vehicle. 'Memorability' was linked to the context factor frequency of use to describe the need for high IVIS usability even after a period of non-use of the device. The final two high-level usability factors, 'satisfaction' and 'perceived usefulness', relate to subjective aspects of usability assessment. Satisfaction was linked to frequency of use and uptake factors as this will affect whether the user chooses to interact with an IVIS, and how frequently. Perceived usefulness relates to users' perceptions of the IVIS, which influence whether they will actually use the device during real driving.

The 12 IVIS usability criteria collectively define usability for IVIS. Each of these criteria is described in more detail by a KPI, which is measurable, either objectively or subjectively. Evaluation against each KPI would enable an evaluator to comprehensively assess the complete usability of any IVIS. KPIs are described according to the attributes of usability they refer to and also in relation to the evaluation which would need to be carried out to measure them. For example, to evaluate the usability of an IVIS under different environment conditions effectiveness needs to be measured under these various conditions. The IVIS usability criteria can be used to guide the design of an IVIS. They can also provide a structure on which to base a comprehensive evaluation of these devices, covering all relevant aspects of usability (Table 2.2).

CONCLUSIONS

A review of the major contributions to defining usability has highlighted the difficulty in developing a general definition of usability. This has led to the conclusion that consideration of the context-of-use is essential in defining the usability criteria for a specific product or system. Ten high-level usability factors were derived from the definitions developed by the leading authors in the field over the last 30 years. In defining usability factors for a particular product, in this case IVIS, the context-of-use must be thoroughly described and analysed. Next, high-level usability factors can be matched and if necessary, redescribed, to reflect the specific context-of-use for which they are intended. Describing context-specific usability factors in this way can help define the boundaries of product design and can contribute to the development of an evaluation process based on these criteria, which is tailored to a particular product. This process was applied to IVIS and a definition of usability, in the form of a list of usability criteria for these devices, has been presented here. KPIs were also identified in order to describe how these criteria can be measured, and these will support the evaluation of IVIS usability. The work presented in this chapter contributes to the understanding of what constitutes usability for IVIS and forms the foundation for the development of an evaluation technique aimed specifically at these products.

3 In-Vehicle Information Systems to Meet the Needs of Drivers

INTRODUCTION

Driving is an example of human–machine interaction in which the human (i.e., the driver) interacts with a machine (i.e., the vehicle). As well as interacting with the primary driving functions, such as steering, accelerating, braking, and changing gear, the driver also performs secondary tasks within the vehicle, and this often involves interacting with an IVIS. To design and evaluate any interactive system, it is necessary to be able to predict how that system will perform, based on the individual components which make up the system. Card, Moran, and Newell (1983) proposed a formula to describe system performance:

Task + User + Computer → System Performance. (1983, p. 404)

The task, user, and computer (or, for the purposes of this work, IVIS) are factors that combine to produce an approximation of system performance. Card et al. (1983) went on to state that modelling the interaction between the task, user, and computer would enable designers to predict system performance:

Model (Task, User, Computer) → Performance Prediction. (1983, p. 405)

These formulae were developed specifically to describe desktop computing systems and are applicable to systems operating in isolation, that is, without reference to the wider context of use. This is a limitation of existing HCI models, such as Card et al.'s Goals, Operators, Methods, and Selection Rules (GOMS) technique, which is underpinned by these formulae for the human–computer system. A major goal of the work described in this book is to account for the context-of-use of a product of system in usability evaluation, and the factors defined by Card et al. must therefore be described in relation to this context, which in this case is interacting with an IVIS whilst driving. Card et al.'s (1983) three factors of a human–computer system are described and discussed in relation to the IVIS interaction/driving context in the following sections of this chapter. An extended version of the human–computer system model is presented in Figure 3.1; this places the interaction between task, user, and

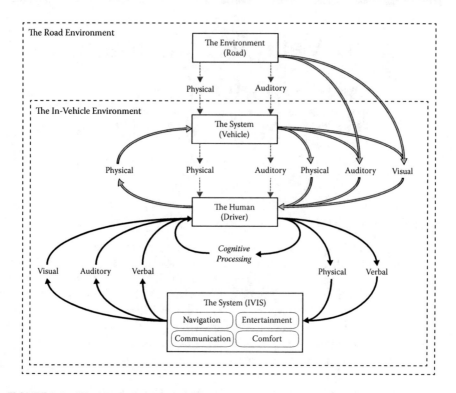

FIGURE 3.1 The interaction between human (driver) and system (IVIS).

system in the IVIS/driving context and shows the modal interactions between the separate components.

THE TASK

Driving is a complex, multitask activity (Regan et al., 2009), consisting of interactions between the driver, the car, and the environment (Rakotonirainy and Tay, 2004) and requiring the successful integration and coordination of the driver's cognitive, physical, sensory, and psychomotor skills (Young et al., 2003). This chapter is concerned primarily with the interaction of the driver with secondary in-vehicle tasks via an IVIS; however, the driver's performance on primary driving tasks is also important, as it can be directly affected by the driver–IVIS interaction.

PRIMARY DRIVING TASKS

During driving, the driver must perform a large number of different tasks while continuously monitoring the driving scene (Wierwille, 1993). Primary driving tasks involve maintaining the safe control of the vehicle (Lansdown, 2000) by guiding its position, detecting and responding to hazards, and navigating a route (Seppelt and Wickens, 2003). Hedlund et al. (2006) listed steering, accelerating, braking, speed choice, lane choice, manoeuvring in traffic, navigation to destination, and scanning for hazards as the primary driving tasks. All primary tasks will be performed by the driver during a single car journey, so it is essential that in carrying out these tasks the driver's performance is not negatively affected.

Secondary (In-Vehicle) Tasks

Secondary tasks are all other tasks performed by the driver that are not directly related to driving (Hedlund et al., 2006). Secondary functions are not essential to successful driving; instead, their purpose is to enhance the driving experience while addressing the driver's needs (Engström et al., 2004; Matthews et al., 2001). Secondary functions provide information about the journey and the vehicle in the form of navigation instructions, traffic information, and vehicle data, which enables the driver to make better informed decisions about that journey and therefore to 'improve the efficiency of roadway use' (Seppelt and Wickens, 2003). They can enhance comfort by enabling the driver to control the climate within the vehicle. Secondary functions provide entertainment, including audio features such as radio, CD, and MP3, and even visual media, including TV and DVD. They can also provide the driver with a means of communication via voice calls and text messages. Traditionally, secondary functions were controlled via hard tactile switches located on the dashboard and centre console. In recent years the number and variety of secondary functions available within vehicles has increased dramatically, from simple radio and climate controls to the vast array of features described above (Gu Ji and Jin, 2010). This has been fuelled by consumer demand for access to more information and enhanced comfort and connectivity while on the move. Today, some lower-end automobile models still use hard switches to control all secondary functions because this is relatively inexpensive. In the premium sector however, and increasingly with volume brands, designers have attempted to integrate many secondary controls into a single menu-based interactive system (Pickering et al., 2007), with only the most high-frequency and high-importance controls left as hard switches. Technologies such as voice recognition and steering wheel-mounted switches are also used as supplementary controls for a number of secondary tasks (Pickering et al., 2007).

THE SYSTEM

IVIS are menu-based systems that enable many secondary functions to be integrated into one system and accessed via a single screen-based interface. This reduces the cluttered appearance of the dashboard. Aesthetically, this approach is superior to the traditional dashboard layout and is ultimately a major selling point for these vehicles: In many cases the IVIS has become a 'brand identifier' (Fleischmann, 2007). IVIS are designed to enhance the driving experience by allowing users to accomplish secondary tasks while driving (Lee et al., 2009). The usability of an IVIS is affected by the HMI, which determines how well a driver can input information, receive and understand outputs, and monitor the state of the systems (Daimon and Kawashima, 1996; Stanton and Salmon, 2009). Although the screen-based interface has improved the visual appeal of the vehicle interior, there are a number of usability issues associated with integrating so many functions into a single system. For example, some functions that could be operated simply via the dashboard are now 'buried' within a complex, multilevel menu structure and require a number of discrete steps to operate (Burnett and Porter, 2001). IVIS present a unique challenge because it is not only the usability of the system that needs to be carefully considered; both (a) the interaction between the IVIS and the primary task of driving, and (b) the potential consequences of this interaction to driving performance and safety are also of vital importance (Dewar et al., 2000). The challenge for designers is to maximize the benefits offered by secondary functions without sacrificing usability and the needs of the driver (Broström et al., 2006; Lee et al., 2009; Walker et al., 2001).

TOUCH SCREENS AND REMOTE CONTROLLERS

Two of the most popular HMI solutions to the challenge of combining a large number of secondary driving controls into a single interactive, screen-based IVIS are the touch screen and remote controller interfaces. The touch screen comprises a screen, located within the zone of reach for the driver, that receives and responds to direct touch from the user. Based on a survey of vehicles from 35 automotive manufacturers, Kern and Schmidt (2009) found that around half of the cars reviewed used a touch screen IVIS. Many British, American, and Japanese manufacturers, including Jaguar Land Rover, Ford, and Toyota, use a touch screen, whilst most German manufacturers, including BMW, Audi, and Mercedes-Benz, prefer controller-based IVIS input (Kern and Schmidt, 2009). The remote controller input type combines a screen, usually placed at the driver's eye level, with a hard control, normally a variation on a traditional rotary dial located on the centre console within reach of the driver. The remote controller is used to navigate through the menus on screen and to select and operate the required functions. Many IVIS also have additional hard buttons, located around the display screen and/or remote controller, to aid menu navigation.

There are a number of features that differentiate the touch screen and remote controller technologies and perhaps explain why neither has emerged as the dominant system in the automotive industry. Input devices can be categorised as *relative* or *absolute* (Douglas and Mithal, 1997), according to their mapping between input parameters and the consequent position of the cursor on the screen, or in other words, whether there is a direct or indirect relationship between the actions of the user on the device and the resulting actions of the device on screen (Douglas and Mithal, 1997; Rogers et al., 2005). Direct devices, of which the touch screen is an example, do not require any translation between the input from the user and the action of the device; in other words there is 'a direct relationship between what the eyes see and what the hands do' (Dul and Weerdmeester, 2001). Indirect devices, on the other hand, do require this translation because the control is remote from the device. The remote controller is an example of an indirect device. The translation of inputs to outputs can be difficult to learn and has been found to hinder performance for novice users of a device (Sutter, 2007). The issue of translation may also present a problem in high-workload situations in which drivers are more likely to make a mistake if what they perceive does not match what they expect (Stevens et al., 2002). On the other hand, indirect devices are considered to be better for high precision and repetitive tasks

(Rogers et al., 2005); however, Rogers et al. (2005) also found that the touch screen was superior for discrete selection tasks and long ballistic movements, and therefore concluded that the specific task requirements must be considered when selecting an appropriate input device. From a physical point of view, the touch screen does not require any associated hard controls and is 'space efficient' (Taveira and Choi, 2009), although the screen must be large enough for each target to be easily distinguishable by touch and prevent accidental activation, and so may need to be larger than the display screen associated with the remote controller. The remote controller can enable higher-precision inputs than the touch screen and can also provide tactile feedback to the user, which can give valuable information about whether they have made the correct input. The lack of tactile feedback afforded by the touch screen is a disadvantage in comparison, although there is evidence of recent work to develop touch screens that provide the user with some form of haptic sensation in response to touch, potentially eliminating this problem from future systems (Graham-Rowe, 2010; Lévesque et al., 2011; Richter et al., 2010). Screens used in combination with a remote controller can be positioned for best visual performance, usually as close as possible to the driver's line of sight. They are also often adjustable and have some level of shrouding to reduce the potential for disabling glare (Howarth, 1991). Touch screens, on the other hand, must be positioned within the zone of comfortable reach (Duland Weerdmeester, 2001) for the driver. This means that the device is often located significantly below the driver's eye line and that any provision of shrouding to protect from glare must be traded off against physical screen accessibility. The position of the touch screen (i.e., so that it is easily visible) may also mean that the driver's arm must be held outstretched during operation, which could result in some level of muscle fatigue (Wang and Trasbot, 2011), and the position of the arm and fingers may mean that part of the screen is obscured (Taveira and Choi, 2009).

Because of the problems discussed above, the increasing use of multifunction, screen-based interfaces by vehicle manufacturers, namely, touch screen and remote controller IVIS, was described as a 'worrying trend' (Burnett and Porter, 2001). Although there has been substantial development in this area in the 10–15 years since these systems were introduced, IVIS are still known to be a significant cause of distraction and are by no means free of usability issues (Regan et al., 2009). There is also concern that technologies such as voice recognition and steering wheel-mounted controls, which are often used to supplement IVIS, have 'inherent limitations without significant safety benefits' (Pickering et al., 2007). There is an obvious need to develop a system that improves usability by overcoming the current problems of the two main IVIS HMIs and other in-vehicle technologies without losing the benefits, in terms of functionality and connectivity, that existing systems currently offer.

THE USER

Today there is a vast array of technologies available to support in-vehicle interactions. In many cases the success of the technology is limited not by the capabilities of that technology but by the capabilities of the human interacting with it. The focus has therefore shifted from development of technology towards consideration of how to integrate this technology with the human element of the interaction—in this case, the driver (Walker et al., 2001). To optimise the human–machine interaction for IVIS, it is important to take a driver-centred approach—in other words, identify and understand the needs of the driver within the context of driving (Heide and Henning, 2006; Stanton and Salmon, 2009). Walker et al. (2001) identified three main driver needs considered to be of importance by automotive manufacturers in relation to the use of information and communication technologies within vehicles: safety, efficiency, and enjoyment. In this context the main aims for an IVIS should be to ensure the safety of vehicle occupants by providing relevant information without distracting the driver from the primary task of driving, to enhance the efficiency of vehicle use by providing information about the vehicle and road network, and to provide functions that are enjoyable to use and enhance the driving experience (Cellario, 2001; Walker et al., 2001). The task for automotive manufacturers is to provide an IVIS that is capable of balancing all three of these driver needs (Tingvall et al., 2009).

SAFETY

Alonso-Ríos et al. (2010) defined user safety as 'the capacity to avoid risk and damage to the user when the system is in use' (2010, p. 61).

On its own, the use of an IVIS poses minimal risk to a user's physical safety; however, when the user–system interaction takes place at the same time as the primary driving task, a driver's safety may be compromised due to the distracting effect of this interaction. This distraction occurs either as a direct result of the functions provided by the IVIS (i.e., loud music) or from the interaction between driver and system (i.e., the driver glancing away from the road to locate functions presented in a visual display; Horrey et al., 2003). Based on results of the 100-Car Naturalistic Driving Study, Klauer et al. (2006) estimated that distraction caused by secondary task interaction contributed to more than 22% of all crashes and near crashes. Hedlund et al. (2006) defined driver distraction as arising from 'any activity that takes a driver's attention away from the task of driving' (2006, p. 1).

This distracting activity can divert attention from the road ahead by creating a mismatch between the attention demanded by the driving environment and the attention the driver is able to devote to it (Lee et al., 2009). If the demands of both the driving environment and the concurrent task are high, then this is likely to exceed the driver's capacity and could lead to distraction (Gu Ji and Jin, 2010; Matthews et al., 2001). There is an upper limit to the level of sensory input a human user can receive and successfully respond to at any time, and automotive manufacturers must therefore balance the provision of information via the IVIS with the capabilities of the human user. This becomes more important as the number of tasks integrated into IVIS increases. The way in which a driver shares attention between competing tasks is very difficult to predict because it is dependent on the immediate situation, specifically the demands of the road environment, and the available capacity of the driver to attend to task performance (Lee et al., 2009). This is influenced by user characteristics and will vary between and within individuals due to factors such as age, experience, stress, and fatigue (Bayly et al., 2009).

Efficiency

The primary aims of efficient driving are to reach the intended destination in an acceptable time, expending a proportionate level of resources (Alonso-Ríos et al., 2010). Some secondary tasks are designed to support the driving task by helping the driver to drive more efficiently (Bayly et al., 2009); for example, navigation information can guide drivers to their destination using the quickest or shortest route possible (Bayly et al., 2009; Walker et al., 2001). Traffic information presented via the radio can also inform the drivers of incidents that could impact on their journey, allowing them to make more informed route decisions and avoid holdups. As well as providing information to the driver, IVIS can also take some control away from the driver by automating tasks in circumstances where technology can offer more efficient performance than the driver (Walker et al., 2001). Automation of tasks can have positive and negative effects on the efficiency of driver performance, and the allocation of functions between the human and system should be given very careful consideration. Drivers also want to be able to interact with the IVIS itself in the most efficient way. This means performing secondary tasks via the IVIS successfully, quickly, with few errors, and within the limits of information-processing capacity (International Organization for Standardization, 1998). The efficiency of the IVIS is determined by the design of the system and its interface. An IVIS with high usability will enable a more efficient interaction between driver and system by presenting clear and useful information to the driver. The driver must also be able to successfully and efficiently input information back to the IVIS and monitor the state of the system for changes.

Enjoyment

For many people driving is not only a means of getting to a destination, it is also an enjoyable experience in itself. Secondary functions can, in some circumstances, relieve the boredom of the driving task and maintain the driver's alertness (Bayly et al., 2009); for example, audio functions offer a source of entertainment to the driver and are aimed at enhancing their enjoyment of driving. Comfort is also an important factor in enjoyment; for example, drivers will not enjoy being in a vehicle that is excessively hot or cold. Enjoyment factors also have an impact on the saleability of vehicles (Walker et al., 2001) and play an important role in brand identification (Fleischmann, 2007; Tingvall et al., 2009). In the competitive automotive market this is essential for consideration in the design and evaluation process. Satisfaction is becoming increasingly important as a factor of the usability of products, as it has a major influence over people's enjoyment of driving. In this context, satisfaction refers to the user's perception of the level of system usability. A system that is perceived to work well, in a way that the user expects, will lead to high levels of user satisfaction (Savoy et al., 2009). Enjoyment is a wider concept, which includes satisfaction but also relates to the functionality of the IVIS and the overall driving experience. Enjoyment can be measured subjectively by evaluating users' preferences; however, Andre and Wickens (1995) argued that the most preferred systems may not always be the best in terms of performance. In designing for usability of IVIS, performance and preference should not, however, be treated as distinct concepts. High usability will enhance a user's interaction with an IVIS, for example, by making it more efficient, effective, and easier to learn. These features of usability are also associated with increasing the user's enjoyment of the interaction, and therefore their preference for the system.

THE TASK-USER-SYSTEM INTERACTION

The individual components in the task-user-system interaction have been defined and discussed in relation to the context-of-use of IVIS. The next step in forming a model to predict and evaluate IVIS usability is to investigate how the task, user, and system interact within a driving context.

MULTIMODAL INTERACTIONS

For tasks to be completed successfully there must be a transfer of information between the user and the system. This usually consists of inputs made by the user to the system and outputs from the system to the user. User inputs can be made via one of two modes: physical, which in the case of most IVIS involves movements such as pushing buttons and turning dials, and verbal, involving the user speaking commands that the system is able to recognize. Secondary driving tasks are controlled primarily by the driver's hands, via the physical mode, although voice-based controls have become increasingly widely used in recent years. System outputs can be made via three different modes: visual, auditory, and physical. The visual mode is the most common mode of information presentation from system to human used while driving (Agah, 2000; Bach et al., 2008; Haslegrave, 1993; Sivak, 1996; Wierwille, 1993), and most IVIS use it as the primary mode of presentation. The auditory mode is relatively underused in driving tasks, in comparison to vision. Use of the auditory mode for secondary task information presentation has received support because auditory tasks can occur simultaneously with visual tasks with minimal interference between the two information-processing channels (Fang et al., 2006; Wickens, 2002). During primary driving, the demands on the auditory mode are also relatively low, and there is spare capacity that could be used in receiving auditory information associated with secondary in-vehicle tasks (Hulse et al., 1998). Compared with the visual mode, physical interaction plays a very small role in information gathering while driving. Haptic feedback, such as vibrations used to alert the driver to new information, is an example of where physical system outputs could be used within a driving environment; however, the range of information and level of detail presented is severely limited in this mode. As well as sending and receiving information to and from a system, the user must also process this information via the cognitive mode. This processing enables the driver to understand the information being presented by the system and make suitable decisions in response to that information. The transfer of information between user and system via these different modes of interaction is illustrated in Figure 3.1. There are interactions between the human (driver), the vehicle, and the IVIS, and these occur within the in-vehicle environment. There are also interactions between the driver, vehicle, and environment, and these occur within the wider road environment. In the diagram, the dashed arrows indicate the transfer of indirect information to the driver, which is transmitted from the road environment via the vehicle to the user. This can include the physical movement of the vehicle as it moves over bumps in the road, which are felt by the driver, and the noise of the tyres as they move over the road. The solid arrows represent direct information: black lines show information involved in primary driving tasks, and grey lines show information involved in secondary IVIS tasks.

TOWARD A PREDICTION OF IVIS USABILITY

The work of Card et al. (1983) showed that it was possible to create models of the task–user–system interaction to enable predictions to be made about system performance. These predictions can then be used to inform system design improvements. Before these system performance predictions can be made, however, it is essential to specify exactly what aspects of system performance are relevant to the particular system under investigation. Defining how the system should perform gives designers and evaluators a benchmark against which to measure actual system performance and decide on the required improvements to design (Gray et al., 1993; Harvey et al., 2011a). In this case the focus was on the usability of IVIS, which influences a driver's interaction with secondary in-vehicle tasks while driving. The context-of-use of these systems was described and defined in the form of six IVIS usability factors: dual task environment, environmental conditions, range of users, training provision, frequency of use, and uptake. This work was the focus of Chapter 2. The IVIS usability criteria must be viewed within the boundaries defined by the needs of the driver to ensure that IVIS are designed to meet these needs and contribute to the overall driving experience. The interaction between driver and IVIS must, in itself, be efficient and must also enhance the efficiency of the complete driving experience. The interaction must also increase, and not detract from, the driver's enjoyment of the driving experience. Finally, these goals of a usable IVIS must not oppose the goal of safe driving.

EVALUATING THE TASK–SYSTEM–USER INTERACTION

This chapter has so far presented descriptions of the system, the users, and their needs, and the type of tasks being performed. The interaction of these three factors has been examined and their relation to usability criteria discussed. This understanding of the user–IVIS interaction provides a base for an evaluation of IVIS. Evaluation involves representing, or modelling, a system and the components of that system, in order to measure the performance of the system. Card et al. (1983) described this system performance as comprising three factors: task, user, and computer. This chapter has discussed the need for these factors to be defined within the correct context-of-use. These components of system performance can be modelled in different ways, and in this study, evaluation is categorised as analytic or empirical. Analytic methods are used to develop models via paper- or computer-based simulations of the interaction, which can be used to predict secondary task performance parameters such as interaction times and potential errors. They generally have low time and resource demands, making them more applicable earlier in the evaluation and design process. It is recommended that analytic methods are applied to predict IVIS usability in order to determine whether a particular design is worth developing further. Systems that are predicted to perform well against the IVIS usability criteria can then be taken forward into the next stage of the framework in which empirical evaluation methods are applied. Empirical methods measure actual performance of human users interacting with system prototypes, generating data on driving and secondary task performance.

The work described so far in this book has demonstrated that usability evaluation is a multistage process, consisting of the specification of a need for the development or redesign of a product or system, an investigation of the context-of-use within which that product or system is used, the definition of usability criteria and KPIs, the description and understanding of factors that determine system performance, and the specification of suitable measurement methods. Over the course of a product development process, it is likely that specific requirements will change, and the availability of resources may improve or reduce. Throughout the process, evaluators' and designers' understanding of the context-of-use and product requirements is also likely to evolve, and it is therefore necessary to take an iterative approach to product development and evaluation. This process is illustrated in Figure 3.2.

This process links the various stages of development, from definition of product or system requirements to final product design, in an iterative cycle. The framework begins with an exploration of the requirements for a product or system, which were defined in this case as usability in order to enhance the driver-IVIS interaction whilst meeting the needs of the driver. The focus of the development/evaluation process is IVIS, in the form of an existing system or a concept design. There is a need to assess the usability of existing IVIS, particularly the touch screen and remote controller interface styles, to support automotive manufacturers in deciding which technology to implement in vehicles. It is also important to evaluate novel IVIS concepts, with a view to improving current interaction styles. Next, in order to understand the challenges of IVIS usability, the context-of-use was defined. This is necessary to highlight wider issues which only emerge when the product is used within a particular situation or environment, or with particular users. The criteria which would need to be met to achieve this goal of a usable IVIS were defined based on knowledge of the product and of the context-of-use. This was done for the context of IVIS to define 12 usability criteria and KPIs (see Chapter 2). These KPIs prescribe the type of information that is needed in order to evaluate the usability of an IVIS and should be used to guide the selection of methods which are most appropriate for evaluating usability. Evaluation involves representing the product and system so that performance can be measured: three factors of system performance, the tasks, users, and computer (or system), need to be defined in order to model the interactions between them and within the wider context-of-use. The knowledge of the product developed via these first four stages of the process then enable suitable evaluation methods to be selected and then applied. In Figure 3.2, methods have been categorised as analytic or empirical. These evaluation stages should then be repeated where necessary in order to refine the design of an IVIS until the usability criteria are met. This iteration

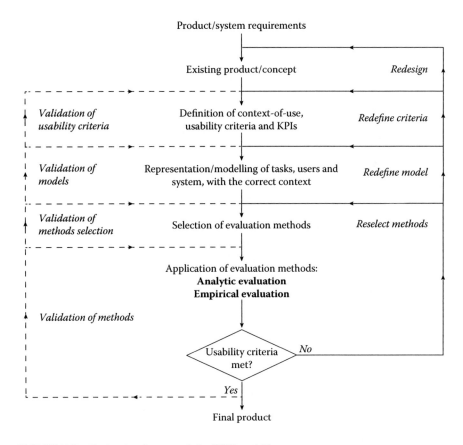

FIGURE 3.2 Evaluation framework for IVIS usability.

validates the findings of each stage of the framework, ensuring that the evaluation process is capable of measuring what it is supposed to measure. This ensures that the results of evaluation are fed back in to the process to inform redesign of the product and that the usability criteria and methods are validated.

CONCLUSIONS

Card et al. (1983) proposed that in order to predict the performance of a particular system, the interaction between three variables—the task, the user, and the system—needs to be modelled. Before this interaction can be modelled, a thorough understanding of the task, user, and system and their interactions within the context-of-use is required, and certain criteria for a target level of system performance must be defined. A multimethod framework was proposed for the evaluation of the usability of IVIS, and this chapter has discussed the information necessary for investigating the task–user–system interaction relating to IVIS. The framework will enable the prediction and measurement of IVIS usability, and the results will be used to inform the redesign of these systems to meet the needs of drivers.

4 A Usability Evaluation Framework for In-Vehicle Information Systems

INTRODUCTION

In Chapter 2, the problem of developing a universal definition of usability was discussed. A universal definition will never capture all of the important factors of usability for every product because consideration of the context-of-use is essential in defining usability criteria, and this will be different for each system under investigation (Harvey et al., 2011a). One of the main purposes of defining criteria for usability is so that it can be evaluated. Usability evaluation is used to assess the extent to which a system's HMI complies with the various usability criteria that are applicable in its specific context-of-use. The results of a usability evaluation can be used to indicate the likely success of a product with its intended market, to compare two or more similar products, to provide feedback to inform design, and even to estimate possible training requirements associated with the product (Butler, 1996; Rennie, 1981).

PREPARING FOR A USABILITY EVALUATION

This chapter describes the development of a usability evaluation framework for IVIS. Before developing the IVIS usability evaluation framework, a number of features relating to this specific system were defined. These related to the interactions which occur between the tasks, users, and system and the context-of-use of IVIS (see Chapter 3). It was also essential to define a comprehensive list of criteria for the usability of IVIS, in order to provide some targets for the evaluation (see Chapter 2). Based on the author's experience of developing the evaluation framework, it is recommended that prior to conducting any usability evaluation, evaluators follow three principles to ensure that important preliminary information is carefully defined: these are presented in Table 4.1.

SELECTING USABILITY EVALUATION METHODS

The success of usability evaluation depends on the appropriateness of the selection of evaluation methods (Annett, 2002; Gelau and Schindhelm, 2010; Kantowitz, 1992). The selection of usability evaluation methods will be a matter of judgement on the part of the evaluator (Annett, 2002), and it is therefore important that he/she has

TABLE 4.1
Three General Principles for Preparing an Evaluation of Usability

Define the task–user–system interaction	These three factors (task, user, system), along with the context-of-use within which they interact, determine the usability of a system and the way in which they will be represented in the evaluation needs to be determined (Harvey et al., 2011b). Unlike the task and the system, the designer has no control over the user of the system; however, the needs of the user and their conceptual model of the interaction must be considered in design (Landauer, 1997; Norman, 2002; Preece et al., 2002; Walker et al., 2001).
Define the context-of-use	The usability of a system is dependent on the context within which it is used (Harvey et al., 2011a). This is because certain attributes of usability will be more or less important depending on the circumstances in which a system is used (Chamorro-Koc et al., 2008; Greenberg and Buxton, 2008). All factors that influence this context-of-use need to be identified.
Define usability criteria	Before a system can be evaluated, evaluators need to know which aspects of the interaction are relevant to usability. Usability criteria, which define a target level of usability, need to be developed.

as much information as possible to inform this choice and to ensure that the evaluation is not weakened by the use of inappropriate methods (Kwahk and Han, 2002; Hornbæk, 2006). Four principles to guide the method selection process were defined following a review of the literature on usability evaluation, according to which many authors advised that consideration of the type of information required, the stage of evaluation, the resources required and people involved is essential in the selection of appropriate methods (Butters and Dixon, 1998; Johnson et al., 1989; Kwahk and Han, 2002; Stanton and Young, 1999b). These four principles, presented and defined in Table 4.2, are closely interrelated, and trade-offs will need to be carefully considered in order to identify appropriate methods in accordance with this guidance.

INFORMATION REQUIREMENTS FOR USABILITY EVALUATIONS

The information required from an evaluation of IVIS usability was defined in the 12 usability criteria and related KPIs presented in Chapter 2, and evaluation methods should be assessed according to their abilities to produce this information. Evaluation methods can be distinguished based on the type of data they deal with, specifically, whether these data are objective or subjective. Objective measures are used to directly evaluate objects and events, whereas subjective measures assess people's perceptions of and attitudes towards these objects and events (Annett, 2002). According to the usability criteria defined for IVIS, a mixture of objective and subjective methods is needed to reflect actual performance levels (e.g., effectiveness, efficiency, and interference), as well as the users' opinions of the IVIS under investigation (e.g., satisfaction and perceived usefulness).

TABLE 4.2
Four General Principles to Guide the Selection of Usability Evaluation Methods

Consider the type of information	The type of data produced by evaluation methods will influence the type of analysis which can be performed (Gelau and Schindhelm, 2010). Interaction times, error rates, user workload and satisfaction are just some of the measures that may be useful in an evaluation and methods should be selected accordingly. A mix of objective and subjective methods are most likely to produce a balanced assessment of usability.
Consider when to test	Evaluation should take place throughout the design process, following an iterative cycle of design-evaluate-redesign (Gould and Lewis, 1985; Hewett,1986; Kontogiannis and Embrey,1997; Liu et al., 2003). Methods should be selected according to their suitability at different stages of design. Methods applied at an appropriate time in the design process should be capable of identifying usability issues before they become too costly to rectify, but without suppressing the development of new ideas (Au et al., 2008; Greenberg and Buxton, 2008; Stanton and Young, 2003).
Consider the resources	The time and resource requirements of a method need to be balanced with the time and resources available for the evaluation (Kirwan, 1992b). Resources include the site of the evaluation, the data collection equipment and the associated costs. Evaluations will also be constrained by the time available and application times should be estimated in order to aid method selection.
Consider the people	The people required for the application of a method will determine its suitability, given the personnel available for the evaluation. Expert evaluators use methods to make predictions about the usability of a system, based on their knowledge and experience (Rennie,1981). Evaluating with users produces measures of the task–user–system interaction and is also useful for investigating subjective aspects of usability (Au et al., 2008; Sweeney et al.,1993). A mix of expert and user tests is recommended to achieve a comprehensive evaluation of usability.

Objective Measures

In an evaluation of usability the objective measures of interest relate to the actual or predicted performance of the system and user during the task-user-system interaction. Objective measures of secondary/primary task interference, such as lateral/longitudinal control and visual behaviour, are affected by the driver's workload, which is likely to be increased during interactions with the IVIS, resulting in decrements in driving control. Objective measures can be used to measure secondary task performance, using data on secondary task interaction times and errors. There are also a number of analytic methods that can predict objective performance data by modelling the task–user–system interaction using paper-based and computer-based simulations.

Subjective Measures

Subjective measures, which involve the assessment of people's attitudes and opinions towards a system, primarily yield qualitative data. Some methods use expert evaluators to identify potential errors, highlight usability issues, and suggest design improvements. The results of these evaluation methods will be determined to some extent by the opinions and prior knowledge of the evaluators involved and may therefore differ between evaluators. The same is true of some subjective, user-based methods, which obtain data on the opinions of a representative sample of users.

WHEN TO APPLY METHODS IN IVIS USABILITY EVALUATIONS

An IVIS will begin as an idea in a designer's mind and may eventually evolve, through various prototype stages, into a complete system. Usability evaluation methods must be appropriate to the stage in the product development process at which they are applied. An iterative process has been suggested for the evaluation of IVIS usability (see the second principle, 'Consider when to test', in Table 4.2). This consists of a cycle of design–evaluate–redesign, which is repeated until the usability criteria are satisfied. In an iterative process, usability evaluation methods should be capable of identifying usability problems at different stages in the process and allowing these problems to be fixed before alterations to design become too costly and time consuming. Methods can be repeated at different stages of the development process to produce new information with which to further refine the design of a system (McClelland, 1991).

Analytic Methods

Analytic methods are used to predict system usability via paper-based and computer-based simulations. They are applicable at any stage of design, providing evaluators have access to a specification of the interaction style and the structure of tasks. It is useful to apply analytic methods as early as possible in the design process so that the predictions they make can inform improvements to the design of IVIS before time and money is spent developing prototype systems (Pettitt et al., 2007; Salvucci et al., 2005).

Empirical Methods

Empirical methods are used to collect data on user performance and workload, under simulated or real-world conditions. They require a much higher level of resources than analytic methods and are not usually applied until later in the design process when initial design problems have been rectified and a prototype system has been developed.

RESOURCES AVAILABLE FOR IVIS USABILITY EVALUATIONS

In order to evaluate the usability of an interface, the task, user, and system need to be represented. The way in which the task, user, and system are represented will be affected by the resources available in an evaluation.

Representing the System and Tasks

The tasks evaluated in any study will be determined by the functionality of the prototype system that is being tested, and this should represent the full range of product attributes of interest in the evaluation (McClelland, 1991). An IVIS can be represented using paper-based design drawings, system specifications, prototypes or complete systems (McClelland, 1991). The level of prototype fidelity can vary dramatically depending on the development stage, product type and features of the product under investigation (McClelland, 1991), and this will affect the validity of the results of the evaluation (Sauer and Sonderegger, 2009; Sauer et al., 2010). The costs associated with product development increase with the level of prototype fidelity so methods which can be used with low specification prototypes, and paper-based or computer-based representations, will be more cost effective (Sauer et al., 2010).

Representing the User

The user can be represented using data generated from previous tests or estimated by an expert evaluator, as is the case with analytic methods. The user can also be represented by a sample of participants who take part in empirical trials. This sample should be representative of the actual user population for the system under investigation. Empirical methods are generally more time consuming than analytic methods because the actual interaction with an IVIS needs to be performed or simulated in real time, with real users. This usually needs to be repeated under different testing conditions. Recruitment of participants and data analysis also imposes high time demands so it may be suitable to use empirical methods to evaluate only a small number of well-developed systems. In contrast, the relative low cost and time demands of analytic methods makes them more suited to evaluating a larger number of less well developed concepts.

The Testing Environment

For empirical methods, the testing environment is also an important factor. Studies of driving performance and behaviour can be conducted in the laboratory or on real roads. In a laboratory-based IVIS usability study, the driving environment is simulated. Driving simulators vary significantly in sophistication from single screen, PC-based systems, to moving base, full-vehicle mock-ups (Gray, 2002; Santos et al., 2005). Simulator studies are valuable for testing users in conditions which may be not be safe or ethical in a real environment (Gray et al., 1993; Stanton et al., 1997). They can also collect a high volume of data in a relatively short time because driving scenarios can be activated on demand, rather than having to wait for certain conditions to occur in the real driving environment (Stanton et al., 1997). Real road studies use instrumented vehicles, equipped with various cameras and sensors, to record driving performance and behaviour. These can be conducted on a test track or on public roads. Real road studies are generally considered to provide the most realistic testing conditions and valid results; however, safety and ethical issues often limit the scope of usability evaluation in these conditions (Santos et al., 2005). In empirical usability evaluations the IVIS also needs to be simulated. The level of system prototype fidelity will be influenced by time and cost constraints, and these limitations must be traded off against the validity of results.

PEOPLE INVOLVED IN IVIS USABILITY EVALUATIONS

As with most systems, the evaluation of IVIS will benefit from testing with both experts and potential users. A framework which included only expert-based methods or only user-based methods could encourage designers to neglect evaluation if the relevant personnel were not readily accessible. The evaluation framework instead allows potential evaluators to select appropriate methods from a wide selection according to the people available in the circumstances.

Usability Evaluation with Users

Involving users in usability evaluation is important for assessing the task-user-system interaction, in particular for identifying the symptoms of usability problems, from which the cause must be identified and rectified (Doubleday et al., 1997). For a user trial, a sample which reflects the characteristics and needs of users, and also the variation in these characteristics and needs, is required (McClelland, 1991; Sauer et al., 2010). The population of potential IVIS users is very large and will include a diverse range of physical, intellectual and perceptual characteristics (Harvey et al., 2011b), which have been described in Chapter 3. These user characteristics must be represented in a valid evaluation of IVIS usability. User trials are generally costly and time consuming, and it can often be difficult to recruit representative samples of adequate size. It is likely that automotive manufacturers will not always have the resources to run extensive user trials, and therefore a supplementary type of evaluation is needed.

Usability Evaluation with Experts

Analytic methods are applied by expert evaluators who aim to identify the causes of usability problems by analysing the structure of tasks and the system interface (Doubleday et al., 1997). This allows predictions about performance and potential usability issues to be made. Evaluators require a certain level of expertise to apply these analytic methods, but for many methods this can normally be gained in a few hours of familiarisation and practice (Stanton and Young, 1999a). The low costs associated with expert evaluations is one of the main advantages of analytic methods, although it is also thought that experts can offer a new and unbiased perspective of a system and are able to provide valuable insights based on their experiences with similar products (Rennie, 1981).

A Flowchart for Method Selection

Four principles for method selection have been defined in discussed in the preceding sections. In order to support analysts in applying these principles, a flowchart to guide method selection was developed: this is organised into sections to represent the four overall principles and includes a series of questions which should be considered by analysts selecting appropriate methods. The flowchart is presented in Figure 4.1. The flowchart is applicable to method selection for the evaluation of any product or system and any context-of-use; however, the steps in the first stage (1. Consider the type of information) ensure that methods are matched with context-of-use and the usability criteria and KPIs which have been defined for this context. The flowchart was used in this study to select methods specifically suited to evaluating IVIS: the selection process and justification for each method choice in terms of the four method selection principles are presented in the following sections.

A Usability Evaluation Framework for In-Vehicle Information Systems

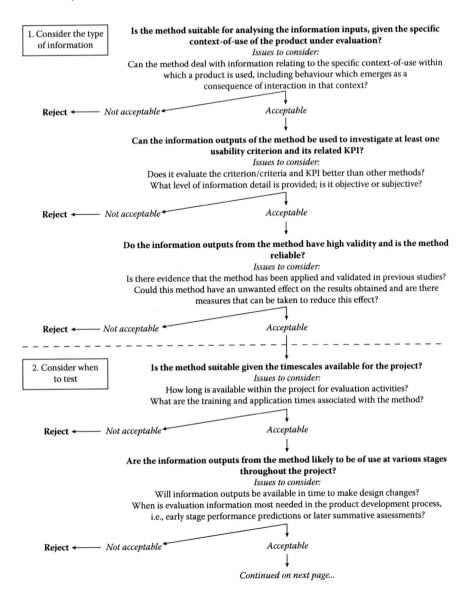

FIGURE 4.1 Flowchart to support the selection of evaluation methods.

USABILITY EVALUATION METHODS

More than 70 usability evaluation methods were identified from the human factors methods literature (see, for example, Karwowski, 2006; Kirwan and Ainsworth, 1992; Nielsen, 1993; Stanton et al., 2005; Stanton and Young, 1999a; Wilson and Corlett, 2005). Methods were reviewed according to the flowchart presented in Figure 4.1 in order to assess their suitability for IVIS usability evaluation. A set of

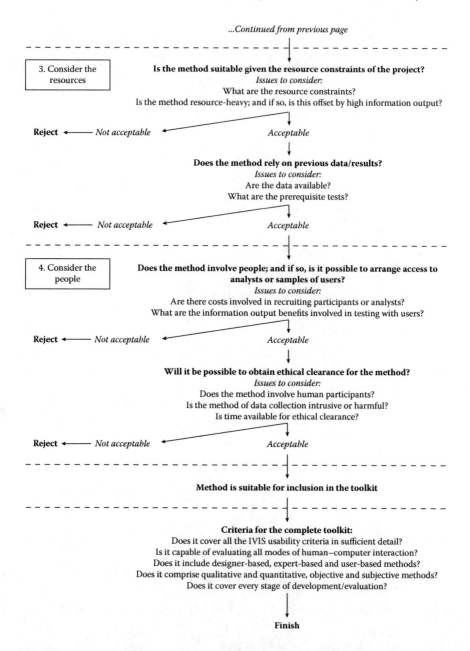

FIGURE 4.1 (continued) Flowchart to support the selection of evaluation methods.

13 methods were selected to make up the final evaluation toolkit as this represented a range of subjective and objective techniques, was applicable at various stages throughout the design process, and covered all the criteria defined for IVIS usability. It is unlikely that one single method will be capable of reflecting a complete picture of the usability of a particular system, and many authors recommend using a range

A Usability Evaluation Framework for In-Vehicle Information Systems

TABLE 4.3
Matrix of Usability Evaluation Methods Matched with IVIS Usability Criteria

IVIS Usability KPIs	Analytic Methods					Empirical Methods						
						Measures of Driving Task Performance				IVIS Task Measures		
	HTA	CPA	SHERPA	Heuristic Analysis	Layout Analysis	Lateral Control	Longitudinal Control	Visual Behaviour	DALI	Secondary Task Times	Secondary Task Errors	SUS
Dual-Task Environment												
1. Task structure and interaction style should be optimal to support IVIS interaction whilst driving	x			x	x							
2. IVIS task times and error rates should be minimised whilst driving		x	x							x	x	
3. Interaction with the IVIS should not affect primary driving				x		x	x	x	x			
Environmental Conditions												
4. Task structure and interaction style should be optimal to support IVIS interaction in all driving conditions					x							
5. IVIS task times and error rates should be minimised in all driving conditions		x	x							x	x	
Training Provision												
6. Task structure and interaction style should be optimal to support IVIS interaction for novice users	x			x	x							
7. IVIS task times and error rates should be minimised for novice users		x	x							x	x	
Range of Users												
8. Task structure and interaction style should support the full range of user characteristics				x						x	x	
Frequency of Use												
9. User satisfaction on initial use and after prolonged use should be high for all aspects of the interaction												x
10. IVIS task times and error rates should be minimised even after a period on non-use of the device										x	x	
Uptake												
11. User satisfaction on initial of the IVIS use should be high for all aspects of the interaction												x
12. Users should report a high likelihood of using the IVIS during real driving												x

of different methods to produce the most comprehensive assessment of usability (e.g., Annett, 2002; Bouchner et al., 2007; Hornbæk, 2006; Kantowitz, 1992). The 13 methods selected for inclusion in the IVIS usability evaluation framework are presented in Table 4.3 in a matrix that matches each method with the usability criteria that it is used to evaluate. Each method is also presented and described in relation to the four method selection principles in the following sections.

ANALYTIC EVALUATION METHODS

Analytic evaluation methods are used to develop symbolic models of the task–user–system interaction via paper-based or computer-based simulations. These models are used to predict IVIS usability parameters such as interaction times and potential errors. Five analytic methods were selected for the IVIS usability evaluation framework: Hierarchical Task Analysis (HTA), Multimodal Critical Path Analysis (CPA), Systematic Human Error Reduction and Prediction Approach (SHERPA), Heuristic Analysis, and Layout Analysis.

Hierarchical Task Analysis (HTA)

HTA is used to produce an exhaustive description of tasks in a hierarchical structure of goals, subgoals, operations and plans (Stanton et al., 2005; Hodgkinson and Crawshaw, 1985). Operations describe the actions performed by people interacting with a system or by the system itself (Stanton, 2006) and plans explain the conditions necessary for these operations (Kirwan and Ainsworth, 1992). HTA is a task

TABLE 4.4
Important Features of HTA

Information	A breakdown of the structure of tasks into individual operations, e.g., move hand to controller, visually locate button. Main use is as a starting point for other methods, although also useful in the assessment of efficiency and effectiveness and to examine if/how tasks are designed to adapt to users.
When to test	Early in the design process, as a precursor to other analytic methods.
Resources	Access to the system under investigation or detailed specification, paper/pen. A relatively time-consuming method (approx. 2–4 hours data collection, 6–8 hours analysis per IVIS), low associated cost.
People	Expert evaluator for data collection and analysis.

analysis method and in most cases needs to be combined with methods of evaluation in order to produce meaningful results (Stanton and Young, 1998b; Stanton, 2006). The important features of HTA are summarised in Table 4.4.

Multimodal Critical Path Analysis (CPA)

CPA is used to model the time taken to perform specific tasks and evaluate how this impacts on other related tasks performed concurrently or subsequently (Baber and Mellor, 2001). For example, it can be used to identify where noncompletion of one task may lead to failure to complete another task (Kirwan and Ainsworth, 1992). CPA is useful for the type of multimodal interactions created by IVIS because, unlike other task modelling methods such as the Keystroke Level Model (KLM) (Kieras, 2001), it can highlight conflicts between primary and secondary tasks occurring in parallel and in the same mode. The important features of CPA are summarised in Table 4.5.

Systematic Human Error Reduction and Prediction Approach (SHERPA)

SHERPA is a human error identification technique designed to identify the types of errors that may occur in performing a task, the consequences of those errors and to generate strategies to prevent or reduce the impact of those errors (Baber and

TABLE 4.5
Important Features of CPA

Information	Predicted task times, modal conflicts, interference from secondary tasks. Task times can be used to assess the efficiency of interaction.
When to test	Useful for predictions of usability at an early stage, although detailed specification of system and tasks is required to produce the initial HTA.
Resources	Access to the system under investigation or detailed specification, database of operation times, paper/pen. A relatively time-consuming method (approx. 2–4 hours data collection, 8–10 hours analysis per IVIS), low associated cost.
People	Expert evaluator for data collection and analysis.

TABLE 4.6
Important Features of SHERPA

Information	Predicted error types, error rates, severity and criticality of errors, error mitigation strategies (i.e., design recommendations). Errors can be used to predict efficiency, effectiveness and interference between primary and secondary tasks.
When to test	Useful for predictions of error at an early stage, although detailed specification of system and tasks is required to produce the initial HTA.
Resources	Access to the system under investigation, database of operation times, paper/pen. A relatively time-consuming method (approx. 2–4 hours data collection, 8–10 hours analysis per IVIS), low associated cost.
People	Expert evaluator for data collection and analysis.

Stanton, 1996; Lyons, 2009). SHERPA can be used to predict where IVIS design issues could cause the driver to make errors in secondary task performance and to develop design improvements specifically to improve aspects of IVIS usability. The important features of SHERPA are summarised in Table 4.6.

Heuristic Analysis

In a heuristic analysis, experts judge aspects of a system or device according to a checklist of principles or 'heuristics' (Cherri et al., 2004; Nielsen, 1993; Stanton et al., 2005; Stanton and Young, 2003). It is usually used to identify usability problems, rather than to assess potential user performance (Burns et al., 2005; Cherri et al., 2004; Jeffries et al., 1991; Nielsen and Phillips, 1993). An advantage of using this checklist approach is that evaluators can be guided towards the aspects of a system that have most influence on usability according to the pre-defined criteria. There are a number of existing checklists and guidelines available for use as part of a heuristic analysis (Alliance of Automobile Manufacturers, 2006; Bhise et al., 2003; Commission of the European Communities, 2008; Green et al., 1994; Japan Automobile Manufacturers Association, 2004; Stevens et al., 1999, 2002; European Conference of Ministers of Transport, 2003), and each has different merits according to the specific features of an evaluation. The principles for method selection should be used to guide the selection of appropriate checklists on a case-by-case basis. The important features of heuristic analysis are summarised in Table 4.7.

Layout Analysis

Layout analysis is a technique used to evaluate an existing interface based on the grouping of related functions (Stanton et al., 2005; Stanton and Young, 2003). It can assist in the restructuring of an interface according to the users' structure of the task. Functions are grouped according to three factors: frequency, importance and sequence of use (Stanton et al., 2005). Layout analysis may be useful in optimising the efficiency of IVIS designs because this will be dependent, in part, on the physical layout of buttons and menu items. The important features of layout analysis are summarised in Table 4.8.

TABLE 4.7
Important Features of Heuristic Analysis

Information	Estimated performance of the system against a list of pre-determined usability criteria, list of usability issues.
When to test	Any stage, although best applied early in the design process to target major usability problems.
Resources	Access to system or prototype, appropriate usability checklist, pen/paper. Relatively low time demands (approx. 1 hr data collection, 1 hr analysis per IVIS), low associated cost.
People	Expert evaluator for data collection and analysis.

TABLE 4.8
Important Features of Layout Analysis

Information	Redesigned layout of menu screens for optimal frequency, importance and sequence of use; number of changes can be used as a quantitative measure. Useful in improving the effectiveness and efficiency of IVIS menu screens.
When to test	Requires knowledge of menu screen layouts from design specifications or existing/prototype systems. Can be used at any stage because only relatively small design changes are identified.
Resources	Access to detailed specifications of menu screens/existing system/prototype, pen/paper. Low-moderate time demands (1–2 hrs data collection, 1 hr analysis per menu screen), low associated cost.
People	Expert evaluator for data collection and analysis.

EMPIRICAL EVALUATION METHODS

Empirical methods measure objective and subjective levels of performance and workload of users interacting with an IVIS. They also evaluate subjective satisfaction and attitudes towards a particular system. In this evaluation framework, empirical methods have been classified as objective or subjective.

Objective Methods

An important criterion for IVIS usability relates to the interference with primary driving caused by interacting with secondary tasks. Primary task performance can be used as a measure of this interference because a driver who is distracted by the IVIS is likely to exhibit degraded driving performance. This degraded driving can be objectively measured by recording lateral and longitudinal control and event detection. Visual behaviour is an objective measure of the proportion of time a driver spends looking at the road compared to the IVIS. Usability can also be evaluated by measuring secondary task performance. This gives an objective measure of the effectiveness and efficiency of the IVIS under investigation. Comparing these objective measures for driving with an IVIS against driving

A Usability Evaluation Framework for In-Vehicle Information Systems

without will indicate the extent to which the usability of the IVIS is interfering with primary driving. Two objective measures of secondary task interaction were selected for inclusion in the framework: secondary task times and secondary task errors. These measures will indicate the effectiveness and efficiency with which secondary tasks can be performed via the IVIS. These measures should be compared across conditions in which the IVIS is used in isolation and simultaneously with the driving task.

Lateral Driving Control

Lateral control is an objective measure which can be used to evaluate the effects of secondary task interaction on primary driving performance (Cherri et al., 2004; Young et al., 2009b). When a driver is distracted from the primary task, particularly by visually demanding secondary tasks, their ability to maintain lateral position on the road is adversely affected (Young et al., 2011a; Young et al., 2009b; Wittmann et al., 2006). The important features of this measure are summarised in Table 4.9.

Longitudinal Driving Control

Longitudinal control is an objective measure relating to the speed of the vehicle (Angell et al., 2006; Cherri et al., 2004; Wittmann et al., 2006). Drivers tend to display greater variations in speed and/or reduced speeds when manually interacting with a secondary task whilst driving (Young et al., 2009b). Longitudinal measures can therefore be used to measure the effect of secondary task interaction on driving performance. The important features of this measure are summarised in Table 4.10.

Visual Behaviour

Visual behaviour can be evaluated by measuring the amount of time the driver's eyes are looking at the road ahead and comparing this to the time spent looking elsewhere (e.g., at the IVIS). This is an objective measure of the interference caused by secondary tasks. If the system is visually distracting, then the driver will spend a significant proportion of the total time looking at it, rather than at the road (Chiang et al., 2004; Noy et al., 2004). The important features of this measure are summarised in Table 4.11.

TABLE 4.9
Important Features of Lateral Driving Control Measures

Information	Lane keeping and steering measures. Poor lateral control would result from interference from secondary task interactions, which could indicate low levels of effectiveness, efficiency, user compatibility, learnability and memorability of the IVIS.
When to test	Relatively late in the design process when access to a full prototype is available.
Resources	Access to a full prototype/complete system, testing environment (lab/real world), test vehicle (simulated/real vehicle), equipment for recording lateral position. High time demands (users are exposed to one or more systems, under one or more testing conditions), relatively high associated cost.
People	Representative sample of user population, experimenters to run user trials.

TABLE 4.10
Important Features of Longitudinal Driving Measures

Information	Speed and following distances. Poor longitudinal control would result from interference from secondary task interactions, which could indicate low levels of effectiveness, efficiency, user compatibility, learnability and memorability of the IVIS.
When to test	Relatively late in the design process when access to a full prototype is available.
Resources	Access to a full prototype/complete system, testing environment (lab/real world), test vehicle (simulated/real vehicle), equipment for recording longitudinal position. High time demands (users are exposed to one or more systems, under one or more testing conditions), relatively high associated cost.
People	Representative sample of user population, experimenters to run user trials.

TABLE 4.11
Important Features of Visual Behaviour Measures

Information	Eyes off road time. This is a measure of the visual distraction caused by secondary tasks.
When to test	Relatively late in the design process when access to a full prototype is available.
Resources	Access to a full prototype/complete system, testing environment (lab/real world), test vehicle (simulated/real vehicle), equipment for tracking driver eye movements. High time demands (users are exposed to one or more systems, under one or more testing conditions), relatively high associated cost.
People	Representative sample of user population, experimenters to run user trials.

Event Detection

A driver's ability to detect and respond to events and hazards in the driving environment can be used as a measure of the interference from secondary tasks (Liu et al., 2009) as it has been shown to be negatively affected by the use of IVIS (Young et al., 2009b; Victor et al., 2009). Event detection can be measured via the number of missed events compared to detected events, the number of incorrect responses to events, the response time, and reaction distance (Young et al., 2009b). The important features of this measure are summarised in Table 4.12.

Secondary Task Times

Monitoring the time a user takes to perform a secondary task gives an objective measure of the time spent away from the primary task, that is, attending to the road ahead. The more time spent on the secondary task, the less time available for attention to driving and therefore the higher the risk to safe driving (Green, 1999). Task time can also be a measure of the effectiveness of the interaction enabled by the IVIS (Noy et al., 2004): the more time required to perform a secondary task, the less effective the interface is. The important features of this measure are summarised in Table 4.13.

TABLE 4.12
Important Features of Event Detection Measures

Information	Number of missed/detected events, incorrect responses, reaction time/distance. This is a measure of the interference from secondary tasks.
When to test	Relatively late in the design process when access to a full prototype is available.
Resources	Access to a full prototype/complete system, testing environment (lab/real world), test vehicle (simulated/real vehicle), equipment for measuring event detection/response time, etc. High time demands (users are exposed to one or more systems, under one or more testing conditions), relatively high associated cost.
People	Representative sample of user population, experimenters to run user trials.

TABLE 4.13
Important Features of Empirical Secondary Task Time Measures

Information	Total task times, individual operation times. These measures can be used to evaluate effectiveness, efficiency, interference, user compatibility, learnability and memorability.
When to test	Relatively late in the design process when access to a full prototype is available.
Resources	Access to a full prototype/complete system, testing environment (lab/real world), test vehicle (simulated/real vehicle), equipment for recording task/operation times. High time demands (users are exposed to one or more systems, under one or more testing conditions), relatively high associated cost.
People	Representative sample of user population, experimenters to run user trials.

Secondary Task Errors

The number and types of errors in the interaction with an IVIS can be used to evaluate the effectiveness of the system design. Errors include pressing incorrect buttons and selecting incorrect functions. Task time compared with number of errors is a useful objective measure of efficiency because it provides information about the quality of the interaction. A usable product will be one which, among other things, enables relatively low task completion times combined with minimum errors. The important features of this measure are summarised in Table 4.14.

Subjective Methods

Subjective methods are used to evaluate the users' perceptions of their primary and secondary task performance and their attitudes towards the IVIS under investigation. Workload can indicate the level of interference caused by interacting with an IVIS. Workload can be measured subjectively, based on self-ratings from users. The level of system usability, with particular reference to user satisfaction, has to be measured subjectively by asking users to rate their experiences with a product.

TABLE 4.14
Important Features of Empirical Secondary Task Error Measures

Information	Number of errors, error types. These measures can be used to evaluate effectiveness, efficiency, user compatibility, learnability and memorability.
When to test	Relatively late in the design process when access to a full prototype is available.
Resources	Access to a full prototype/complete system, testing environment (lab/real world), test vehicle (simulated/real vehicle), equipment/observer for recording errors. High time demands (users are exposed to one or more systems, under one or more testing conditions), relatively high associated cost.
People	Representative sample of user population, experimenters to run user trials.

Driving Activity Load Index (DALI)

DALI is a method for measuring users' subjective workload. It is based on the NASA-TLX workload measurement scale and is designed specifically for the driving context (Pauzié, 2008). Unlike NASA-TLX, DALI includes a rating for interference with the driver's state caused by interaction with a supplementary task. Participants are asked to rate the task, post-trial, along six rating scales: effort of attention, visual demand, auditory demand, temporal demand, interference and situational stress (Johansson et al., 2004). The DALI questionnaire is shown in Figure 4.2. The important features of DALI are summarised in Table 4.15.

System Usability Scale (SUS)

SUS is a subjective method of evaluating users' attitudes towards a system, consisting of 10 statements against which participants rate their level of agreement on a 5-point Likert scale (Brooke, 1996). The SUS questionnaire is shown in Figure 4.3. A single usability score is computed from the ratings and this is used to compare participants' views of different systems (Bangor et al., 2008). Examples of statements include 'I needed to learn a lot of things before I could get going with this system' and 'I think I would like to use this system frequently'. These cover two of the criteria for IVIS usability: learnability and satisfaction (Brooke, 1996). SUS is applicable to a wide range of interface technologies and is a quick and easy method to use (Bangor et al., 2008). The important features of SUS are summarised in Table 4.16.

CONCLUSIONS

The main aim of chapter was to present a flowchart to guide the selection of usability evaluation methods and to use this to select a set of analytic and empirical methods which are suitable for the evaluation of IVIS. A literature review to explore the method selection process was conducted as part of this work and four general principles for method selection were identified. These have been presented here as a useful guide to selecting appropriate evaluation methods according to the type of information required, the stage of application, the resources available, and the personnel

A Usability Evaluation Framework for In-Vehicle Information Systems

Driving Activity Load Index (DALI) Questionnaire

During the test you have just completed you may have experienced some difficulties and constraints with regard to the driving task.

You will be asked to evaluate this experience through 7 different factors, which are described below. Please read each factor and its description carefully and ask the experimenter to explain anything you do not fully understand.

Factor	Description
Global attention demand	Mental (i.e. to think about, to decide...), visual and auditory demand required during the test to perform the whole activity
Visual demand	Visual demand required during the test to perform the whole activity
Auditory demand	Auditory demand required during the test to perform the whole activity
Tactile demand	Specific constraints induced by tactile vibrations during the test
Stress	Level of stress (i.e. fatigue, insecurity, irritation, feelings of discouragement) during the whole activity
Temporal demand	Pressure and specific constraint felt due to time pressure of completing tasks during the whole activity
Interference	Disturbance to the driving task when completing supplementary tasks via the IVIS simultaneously

For each factor you will be required to rate the level of constraint felt during the test on a scale from 0 (very low level of constraint) to 5 (very high level of constraint), with regard to the driving task

Global attention demand:

Think about the mental (i.e. to think about, to decide...), visual and auditory demand required during the test to perfom the whole activity

Low   High

0   1   2   3   4   5

☐   ☐   ☐   ☐   ☐   ☐

Visual demand:

Think about the visual demand required during the test to perform the whole activity

Low   High

0   1   2   3   4   5

☐   ☐   ☐   ☐   ☐   ☐

FIGURE 4.2 Driving Activity Load Index (DALI) questionnaire (Pauzié, 2008; reproduced with kind permission from the author).

Auditory demand:

Think about the auditory demand required during the test to perform the whole activity

Low   High

0   1   2   3   4   5

☐   ☐   ☐   ☐   ☐   ☐

Tactile demand:

Think about the specific constraints induced by tactile vibrations during the test

Low   High

0   1   2   3   4   5

☐   ☐   ☐   ☐   ☐   ☐

Stress:

Think about your level of stress (i.e. fatigue, insecurity, irritation, feelings of discouragement) during the whole activity

Low   High

0   1   2   3   4   5

☐   ☐   ☐   ☐   ☐   ☐

Temporal demand:

Think about the specific constraints felt due to time pressure of completing tasks during the whole activity

Low   High

0   1   2   3   4   5

☐   ☐   ☐   ☐   ☐   ☐

Interference:

Think about the disturbance to the driving task when completing supplementary tasks (i.e. via the IVIS) simultaneously

Low   High

0   1   2   3   4   5

☐   ☐   ☐   ☐   ☐   ☐

FIGURE 4.2 (continued) Driving Activity Load Index (DALI) questionnaire (Pauzié, 2008; reproduced with kind permission from the author).

TABLE 4.15
Important Features of DALI

Information	Users' subjective ratings of six aspects of perceived workload. Workload can indicate the effectiveness and efficiency of task performance, primary/secondary task interference, compatibility of the system with different users, and learnability.
When to test	Relatively late in development when access to a full prototype is available.
Resources	Access to a full prototype/complete system, testing environment (lab/real world), test vehicle (simulated/real vehicle), questionnaire, recording material. High time demands (users are exposed to one or more systems, under one or more testing condition, then need to answer the questionnaire for each condition), relatively high associated cost.
People	Representative sample of the user population, experimenters to run user trials and administer the questionnaire.

involved in the evaluation. These principles were used to structure a flowchart to support the method selection process. The flowchart was then used to select 13 methods which were appropriate for evaluation in an IVIS context. These 13 evaluation methods have been presented and discussed. Five of the methods in the framework were applied in an analytic evaluation of two IVIS, and the results of this evaluation are presented in Chapter 5. An empirical evaluation of two IVIS was also conducted, using the remaining methods in the framework: this study is described in Chapter 6.

System Usability Scale (SUS) Questionnaire

We would like to ask you about the usability of the in-vehicle device you just used.

Please rate how much you agree with the following statements (please tick one box per row):

	Strongly disagree 0	1	2	3	4	Strongly agree 5
I think I would like to use this system frequently	☐	☐	☐	☐	☐	☐
I found the system unnecessarily complex	☐	☐	☐	☐	☐	☐
I thought the system was easy to use	☐	☐	☐	☐	☐	☐
I think I would need the support of a technical person to be able to use this system	☐	☐	☐	☐	☐	☐
I thought the various functions in this system were well integrated	☐	☐	☐	☐	☐	☐
I thought there was too much inconsistency in this system	☐	☐	☐	☐	☐	☐
I would imagine that most people would learn to use this system very quickly	☐	☐	☐	☐	☐	☐
I found this system very awkward to use	☐	☐	☐	☐	☐	☐
I felt confident using the system	☐	☐	☐	☐	☐	☐
I needed to learn a lot of things before I could get going with the system	☐	☐	☐	☐	☐	☐

FIGURE 4.3 System Usability Scale (SUS) questionnaire (Brooke, 1996; reproduced with kind permission from the author)

TABLE 4.16
Important Features of SUS

Information	Users' subjective ratings of ten aspects of system usability. The ten SUS rating scales cover many aspects of the usability of IVIS and are particularly useful in addressing the issue of uptake, which can only be evaluated subjectively.
When to test	Mid-late in the design process when access to a part/full prototype is available.
Resources	Access to a part/full prototype/complete system, questionnaire, recording materials. SUS can be used to evaluate an IVIS in isolation or situated in the vehicle. May also require testing environment (lab/real world), test vehicle (simulated/real vehicle). Medium-high time demands (depending on the test set-up, users may be exposed to one or more systems, under one or more testing conditions, then answer the questionnaire for each condition), relatively high associated cost.
People	Representative sample of user population, experimenters to run user trials and administer questionnaire.

5 The Trade-Off between Context and Objectivity in an Analytic Evaluation of In-Vehicle Interfaces

INTRODUCTION

This chapter presents a case study to explore the use of analytic methods in the IVIS development cycle. The motivation for the work was to understand how to deliver an approach to modelling aspects of IVIS usability, working with inevitable commercial constraints, to provide useful information on which to base design decisions. Analytic methods were selected to meet a requirement for an approach to evaluation which can be applied at an early stage of product development with little demand for resources; however, currently these methods are not widely used in the automotive industry for IVIS evaluation. In this chapter we therefore attempt to explore the utility of analytic methods, including advantages and disadvantages, identify training and application times, and address shortcomings by proposing extensions to one or more of the techniques to increase their utility in a driving context. The findings will be useful to interface designers and evaluators working within the automotive industry, but also in other domains, to support the selection and application of analytic methods, with the overall objective of encouraging early-stage evaluation and design for usability. An overview of the procedures for carrying out the methods is also provided for students, researchers and engineers who are relatively new to the techniques.

The case study was used to evaluate two IVIS: a touch screen, which is one of the most commonly used interface types in current vehicles, and a remote joystick controller, which works like a joystick to control a cursor on screen and was recently introduced to the market. It is important for automotive manufacturers to evaluate the performance of a new IVIS interface technology like the joystick controller against their current system, as a benchmarking activity. The results of this comparison are reported in the case study; however, the main aim was to explore the intrinsic attributes of analytic methods in the context of IVIS evaluation (Gray and Salzman, 1998), rather than as a direct comparison of systems.

ANALYTIC METHODS

Analytic methods were selected to measure various aspects of system performance in order to evaluate IVIS against the KPIs defined in Chapter 2. The KPIs addressed by the set of analytic methods applied in this case study are shown in methods matrix in Chapter 4 (Table 4.3). Today, usability evaluation is widely encouraged in academia and industry; however, there have been suggestions that it can be ineffective and even detrimental if applied blindly and according to rule, rather than as a method of encouraging thought and consideration in designers and developers (Greenberg and Buxton, 2008). Automotive manufacturers also tend to employ two distinct approaches to IVIS evaluation: driving performance measured in relation to safety of driving whilst using an IVIS, and customer satisfaction measured by surveys (Broström et al., 2011). The analytic methods presented in this study were selected to meet a requirement for measures that give an indication of interface usability before a product is sent to market and that encourage designers to explore how the design of an interface influences the user experience. A review of analytic methods was conducted and the five methods presented in this study were identified as most suitable in an IVIS context, given the constraints of the automotive industry described above: this selection followed the process outlined in the flowchart in Chapter 4.

An overview of the five analytic methods is presented in Table 5.1. This also identified which KPIs the methods are used to assess; the numbers correspond to the KPIs listed in Chapter 2 (Table 2.1). Table 5.2 lists the inputs and outputs of each method. Heuristic analysis and layout analysis yield mainly qualitative data; CPA and HTA are used to generate mainly quantitative data, and SHERPA produces both quantitative (error rate) and qualitative (remedial strategies) information (Stanton and Young, 1999a).

METHOD

An evaluation of two existing IVIS was performed using the five analytic methods in order to explore the utility of this approach, in terms of information inputs and outputs, training times, resource demands, and possible extensions, in the context of early-stage product development.

Equipment

The IVIS

The systems under investigation were a touch screen IVIS and a remote joystick controller IVIS. Both IVIS were manufacturer-installed systems, situated within their respective vehicles. The systems used their own manufacturer-designed Graphical User Interfaces (GUI), which meant that the layout of icons was different for the two IVIS. Figure 5.1 illustrates the typical layout of these systems, showing the position of the display screen and additional control pad (this was only present in the remote controller system). The touch screen IVIS was located to the left of the driver on the

TABLE 5.1
Analytic Methods and Related KPIs

KPIs	Analytical Methods	Description
1, 6	HTA	A task description method, used to break down tasks into their smallest components and structure them in a hierarchy of goals, sub-goals and plans (Kirwan and Ainsworth, 1992; Stanton et al., 2005). Although HTA normally needs to be combined with other techniques to produce meaningful analysis (Stanton, 2006), it can illustrate where tasks might lead to ineffective interactions due to poor structure.
2, 5, 7	CPA	Used to model task times based on the interactions between operations performed in different modes (Baber and Mellor, 2001; Wickens, 1991). CPA was selected over other time-prediction methods as it enables operations to be modelled in parallel. Task times produce high correlations with eyes-off road time (Green, 1999; Nowakowski et al., 2000), which is a measure of the interference of secondary tasks in the dual task environment.
2, 5, 7	SHERPA	Predicts error rates and types for particular systems and tasks (BaberandStanton, 1996; Lyons, 2009). Errors will be useful in assessing the level of training which is needed for successful use of a product or system. The nature of the dual task driving environment will also give rise to specific errors, such as failing to complete an operation due to a sudden increase in primary task demand.
1, 3, 4, 6, 8	Heuristic Analysis	Uses a checklist of principles as a guide for identifying usability issues with an interface (Nielsen, 1993). The content of the analysis is set according to the criteria of interest: dual task environment, environmental conditions, range of users and training provision. Because it is a subjective technique, it is less easy to predict factors such as uptake, which needs to be evaluated with real users.
1, 6	Layout Analysis	A method for evaluating an interface based on the position of related functions, according to frequency, sequence, and importance of use (Stanton et al., 2005). It is related to the dual task criterion because the location of an IVIS in relation to the driver will affect the optimisation of layout. It is also related to frequency of *use* because familiarity of users with the interface is a factor that determines layout.

vehicle dashboard and the interaction involved touching targets directly on screen. In the remote controller vehicle, the joystick controller was situated low down in the vehicle's centre console. This controlled actions on a screen which was recessed into the dashboard, to the left of the driver. In the remote joystick vehicle there was also a hard 'enter' button located on the side of the controller unit.

TABLE 5.2
Inputs and Outputs for Analytic Methods

Method	Inputs	Quantitative Outputs	Qualitative Outputs
HTA	Task specification	Number of operations, hierarchical task structure	Understanding of task, goals and plans
CPA	HTA	Task interaction times	Operation dependencies
SHERPA	HTA	Error types and frequencies	Remedial strategies
Heuristic Analysis	Experience of system/task specification	Number of usability issues identified	Types of usability issues, potential problems
Layout Analysis	System layout diagrams	Number of layout changes required	Changes to interface layout

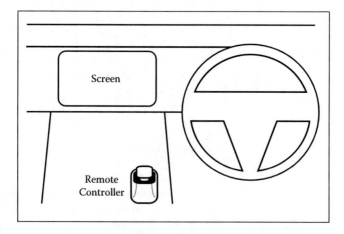

FIGURE 5.1 Schematic showing the relative positions of the IVIS screen and joystick controller.

Data Collection Apparatus

The equipment required for data collection included paper copies of the checklist, paper for recording observations, a camera for taking pictures of the systems, and a sound recorder for capturing audio information. The checklist used in the heuristic evaluation was developed by Stevens et al. (1999) and was adapted for this evaluation by removing sections which were not directly connected to usability. The checklist has recently been updated to reflect developments in IVIS Ergonomics and safety issues. An extract from the checklist is shown in Figure 5.2. (See Stevens et al. (2011).

PROCEDURE

Expert walkthroughs of two existing IVIS were performed by a Human Factors analyst. These were based around a scenario of interacting with several in-vehicle,

Trade-Off between Context and Objectivity in an Analytic Evaluation

```
B3 Is physical or visual access to the IVIS free from obstruction by other driver controls/ displays?
        No IVIS displays are obstructed              TRUE/FALSE/NA
        No IVIS controls are obstructed              TRUE/FALSE/NA
  None  □        Minor  □        Serious  □      NA  □

B4 Is the IVIS free from reflections and glare under all ambient lighting conditions?
        The IVIS is free from reflection/glare:
                during the day                        TRUE/FALSE/NA
                during darkness                       TRUE/FALSE/NA
  None  □        Minor  □        Serious  □      NA  □
```

FIGURE 5.2 Extract from the Stevens et al. (1999) checklist for IVIS usability. (Reproduced with kind permission of the first author.) An extract from the checklist is shown in Figure 5.2.

secondary tasks in a stationary vehicle. The term 'task' is used to refer to a sequence of operations performed by a user to achieve a goal, such as selecting a radio station or reducing fan speed. A single analyst applied all five methods reported in this study. Heuristic Analysis was performed first, whilst the analyst was interacting with each interface; the other four methods were applied after the data collection phase using the information gathered from each IVIS, in the order HTA, CPA, SHERPA, and Layout Analysis. The analyst trained in each of the methods prior to the data collection phase and spent approximately 4–5 hours using the two IVIS interfaces before applying the methods. The analyst had extensive background knowledge of Ergonomics, specifically user–vehicle interactions and experience with Human Factors methods for IVIS evaluation. This single analyst approach is typical of IVIS usability evaluation in industry, which is often subject to tight time and resource constraints. A set of nine representative IVIS tasks was defined for this study, as shown in Table 5.3.

TABLE 5.3
IVIS Tasks Analysed in the Evaluation

Categories	Tasks
Audio	Play radio station: 909 am (radio is currently set to 97.9 fm)
	Increase bass by two steps
Climate	Increase temperature by 1°C (via centre console controls, not IVIS)
	Reduce fan speed by two steps
	Direct air to face only (air direction is currently set to face and feet)
	Direct air to face and feet (air direction is currently set to windscreen only)
	Activate auto climate (via centre console controls, not IVIS)
Navigation	Set navigation from system memory: 'Home'
	Set navigation from previous destinations: 'University Road, Southampton'

These nine tasks were selected from a set of more than 130 tasks which were identified for existing IVIS from a review of automotive manufacturers' IVIS manuals, which was conducted by the authors prior to data collection. Four factors, defined by Nowakowski and Green (2001), were used to guide task selection: use whilst driving, availability in existing systems, frequency of interaction, and compliance with the 15-second task time rule (see Green, 1999). The nine tasks were all likely to be used whilst driving, unlike other functions such as vehicle or display settings. Based on information from automotive manufacturers and the authors' personal experience, it was expected that the tasks would all be used at least once during a typical medium-long journey. All of the tasks were available in existing IVIS, including the two systems under investigation. It was important that the tasks were available in both systems so that valid comparisons could be made. Many tasks differed in structure between the two systems and this limited the choice of suitable tasks for the evaluation. The tasks were all expected to be used fairly frequently by drivers, based on information from manufacturers and the literature. Finally, preliminary investigations indicated that it should be possible to complete each of the nine tasks in less than 15 seconds. The 15-second rule, which is commonly referred to in the design and evaluation of IVIS tasks (Green, 1999; Nowakowski and Green, 2001; Society of Automotive Engineers, 2002), states that no navigation tasks involving a visual display and manual controls, and available during driving, should exceed 15 seconds in duration (Green, 1999). Tasks were performed using each system and the inputs (from user to system) and outputs (from system to user) were recorded. Pictures were taken of the IVIS menu screens and controls at each stage of the interaction and the analyst recorded a description of each interaction. For the Heuristic Analysis, each IVIS was assessed against the Safety Checklist for the Assessment of In-Vehicle Information Systems (Stevens et al., 1999). This checklist was developed by the Transport Research Laboratory (TRL) for the UK's Department of the Environment, Transport and the Regions (DETR), to assess new IVIS. Guidelines to support the checklist recommend that the checklist should only be used to assess functions which are present in a system, and consequently some sections which were not directly connected to usability were removed. These included sections relating to the documentation supplied with an IVIS, the packaging of the product, compliance with traffic regulations, system maintenance, and information referring to the road network.

DATA ANALYSIS

The data collected on each IVIS was modelled using the five analytical evaluation methods described previously. During the modelling phase, close attention was paid to the utility of each method and to the training times, execution times and resources required. Each of the methods was described in Chapter 4 and further details are given in the following sections where necessary. The results of the application of each method and the utility of each technique are also described.

RESULTS AND DISCUSSION

HIERARCHICAL TASK ANALYSIS (HTA)

HTAs were developed for the two IVIS under investigation. An overall goal was defined for each task: this was divided into sub-goals, which were broken down further into operations, for example, the smallest level of activity which makes up the task (Stanton and Young, 1999a). When each task was dissected to the level of operations, plans were generated to describe how the operations are performed to achieve the task goal. Each HTA was recorded as a hierarchical, numbered list. An example of a HTA for the remote controller IVIS task, 'play radio station' is presented in Figure 5.3.

HTA for IVIS Evaluation

HTA is a task description method. Task description is a necessary precursor for further analysis, such as CPA, which will produce measurable results (Stanton and Young, 1999a). HTAs for two or more systems may be subjectively compared in order to identify differences in task structure; however, this exercise is useful for task exploration, rather than as a method for contrasting products. It is possible to compare two or more different products or individual tasks using number of operations, as identified by HTA. The tasks analysed in this case study consisted of a total of 142 individual operations for the touch screen system and 113 operations for the remote controller: these values are broken down by task in Table 5.4. A system which requires the user to perform a large number of operations per task is likely to be less efficient than a system with fewer operations; however, this will also depend on the time taken to perform each operation and the error potential of the tasks involved.

Five out of the nine tasks required more operations to complete with the touch screen than the remote controller and two of the tasks required more operations with the remote controller than the touch screen. Two tasks, 'Increase temperature' and 'Auto climate', required the same number of operations: this was because they were operated using centre console controls and the task structure was the same in both cases. The differences between the results were initially investigated by examining the individual task segments which were common to all tasks: these represent the selection of a single target and are the basic 'building blocks' of all menu navigation tasks. A menu selection task segment for the touch screen consists of four operations: 'Make selection', 'Locate target', 'Move hand to target', and 'Touch target'. The same task segment for the remote controller also consists of four operations: 'Make selection', 'Locate target', 'Move pointer to target' and 'Press enter button'. Because the task segments for the touch screen and remote controller consist of the same number of individual operations, there must be another reason for the difference in total number of operations between the two systems: task structure. Task structure describes the method by which a user completes a task, in terms of the menu target selection required on each menu screen. For example, the 'Enter destination from system memory' touch screen task requires the user to read and accept a navigation warning

1 Play radio station
Plan 1 - Do 1—3 in order to select station; THEN 4 to confirm selection

    1.1 Open AUDIO menu
    Plan 1.1—Do 1 and 2 together; THEN 3, WHEN cursor is over target THEN 4

        1.1.1 Move hand to controller
        1.1.2 Prepare to open menu
        Plan 1.1.2 - Do 1 and 2 together

            1.1.2.1 Make selection
            1.1.2.2 Locate AUDIO icon

        1.1.3 Move point to AUDIO icon
        1.1.4 Press enter button

    1.2 Open AM tab
    Plan 1.2—Do 1, 2; WHEN cursor is over target THEN 3

        1.2.1 Prepare to open tab
        Plan 1.2.1 - Do 1 and 2 together

            1.2.1.1 Make selection
            1.2.1.2 Locate AM tab

        1.2.2 Move pointer to AM tab
        1.2.3 Press enter button

    1.3 Select 909 AM station
    Plan 1.3—Do 1, 2; WHEN cursor is over target THEN 3

        1.3.1 Prepare to select station
        Plan 1.3.1 - Do 1 and 2 together

            1.3.1.1 Make selection
            1.3.1.2 Locate 909 AM button

        1.3.2 Move point to 909 AM button
        1.3.3 Press enter button

    1.4 Confirm selection
    Plan 1.4—Do 1 to confirm correct selection; IF correct THEN 2

        1.4.1 Confirm selection
        Plan 1.4.1 - Do 1 and 2 together

            1.4.1.1 Check 909 AM button is highlighted
            1.4.1.2 Listen to check selection

        1.4.2 Replace hand on steering wheel

FIGURE 5.3 Excerpt of HTA for 'play radio station' task performed using the remote controller IVIS.

TABLE 5.4
Number of Operations in Each Task for the Two IVIS, According to HTA

Task	Touch Screen	Remote Controller
Play radio station (909AM)	15	16
Increase bass by 2 steps	19	16
Increase temperature by 1 degree	6	6
Reduce fan speed by 2 steps	10	11
Direct air to face and feet	18	10
Direct air to face only	14	10
Turn on auto climate	6	6
Enter destination from system memory	27	19
Enter destination from previous entries	27	19
Total operations	**142**	**113**

and to 'start' the route guidance after the destination has been entered. The same task with the remote controller did not require these extra task segments; therefore, there were fewer total operations required to complete the task. This information indicates to a designer where effort in redesign needs to be concentrated, that is, task structure, in order to minimise the number of operations needed to perform a task. In the case of the navigation warning, automotive manufacturers need to consider the trade-off between providing safety-related warnings to drivers about the risks of interacting with the navigation system whilst driving, against the extra time that this adds to the task of programming the navigation system.

HTA Utility

HTA is a fairly time-consuming method to carry out as each individual operation in a task needs to be analysed; however, creating a comprehensive HTA can considerably reduce the time required for other modelling methods such as CPA and SHERPA. A problem facing HCI is that interfaces are often engineering-focussed and are therefore not optimised for activity patterns (Wilson, 2006). HTA provides an activity-based classification of user behaviour, which in itself can be used to improve interface design. The process of conducting HTA can also provide the analyst with important information about task structure and menu design. In this study HTA highlighted the different operations involved in the two IVIS; for example, the structure of tasks resulted in a larger number of operations for the touch screen than the remote controller. This information is useful in the refinement of task design and for understanding the causes of difference in performance between IVIS. A deeper understanding of the links between task design and usability should increase focus on good HMI design. It is therefore also recommend that designers and Ergonomics specialists within manufacturing companies use the process as a learning tool.

MULTIMODAL CRITICAL PATH ANALYSIS (CPA)

CPA is a method that is used to model the time taken to perform specific operations to produce a prediction of total task time (Baber and Mellor, 2001; Harrison, 1997; Lockyer, 1984). The technique was originally developed for project analysis to support the planning and control of work activities (Harrison, 1997), and more recently it has been applied in the analysis of human response times (e.g., Stanton and Baber, 2008). In CPA, a task is divided into operations which have a definable beginning and end (Harrison, 1997), that is, 'visually locate target', and 'touch target'. These operations are categorised as visual, manual, cognitive, or auditory. Operations can be identified by HTA, which divides tasks into the smallest possible levels of activity.

Operations occur in series or in parallel to make up a complete task. Parallel activities can be described according to the Multiple Resource Model (Wickens, 2002). This theory proposes that attention can be time-shared more effectively between operations across different modes, compared with operations which utilise the same mode (Wickens, 2002). Two visual operations—for example, locating a control on the vehicle's dashboard and reading a label on screen—cannot occur in parallel; however, one of these visual operations may take place at the same time as a physical operation, such as moving the hand towards the screen. An advantage of CPA over other network analysis methods such as the KLM, is that it is capable of modelling parallel as well as serial operations. The structure of a CPA model is also affected by the dependency of operations. A dependent operation is one which cannot begin until a previous operation has been completed (Baber and Mellor, 2001; Baber, 2005a). For example, the driver cannot press the enter key on a remote controller until the pointer has been moved to the on-screen target. Each operation is represented pictorially as a node, and the relationships between the operations are denoted by connecting arrows and their relative positions in the CPA diagram (Harrison, 1997), as shown in the example in Figure 5.4.

Defining CPA Activities

In the CPA diagram, time flows from left to right; therefore, a succeeding activity that is dependent on the completion of a preceding operation is positioned to the right of the operation upon which it is dependent. Parallel operations are located in the same vertical position in the diagram and are separated into rows to represent the different interaction modes (visual, manual, cognitive, and auditory). After modalities and dependencies are defined, durations can be assigned to each operation. In this study, these operation duration times were derived from a review of the HCI literature and are listed, along with their sources, in Table 5.5. There are also a number of rules and assumptions that support the use of these timings in the CPA models:

- Time to visually locate a target is 1300 ms, following Stanton and Baber (2008), for any single target and the first alphanumeric target in a sequence.
- Time to visually locate a target is 340 ms for any sequential alphanumeric target after the first target in a sequence. It is assumed that users would be more familiar with the layout of an alphanumeric keyboard than with the other menu screens in each system; therefore, search time for alphanumeric

Trade-Off between Context and Objectivity in an Analytic Evaluation

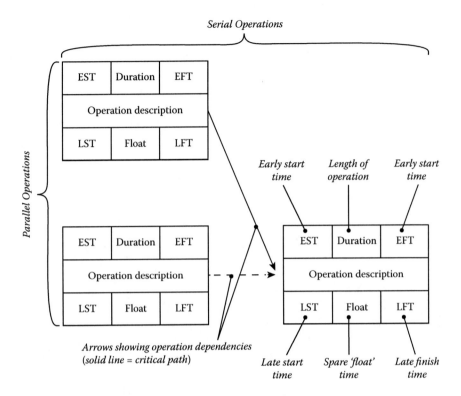

FIGURE 5.4 CPA nodes and dependency arrows in a network diagram.

targets was reduced to 340 ms, following the time to recognize familiar objects reported by Olson and Olson (1990).
- No cognitive 'make selection' operation occurs in parallel with a sequential alphanumeric visual search (340 ms), following the heuristics for Mental operators devised by Card et al. (1983). Entering a word or telephone number into the system is assumed to be a single 'chunk': users make a decision about the sequence of letters or numbers at the start of the chunk; therefore, individual decisions for each alphanumeric are assumed to be unnecessary.
- There is always some movement of the hand/fingers (touch screen) or the cursor (remote controller) during visual search. This movement follows the direction of gaze so only a small 'homing' movement is needed when the target is found (Olson and Olson, 1990). This movement time is not fixed as it varies with the visual search time. It is assumed that the movement starts just after visual search begins; therefore, a value of 1000 ms has been assigned in the models.

Duration, modality and dependency information is used to calculate Early Start Time (EST) and Early Finish Time (EFT) as part of the forward pass through the network; Late Start Time (LST) and Late Finish Time (LFT) as part of the backward pass through the network, and finally, Float Time, according to the following rules:

TABLE 5.5
Operation Times from HCI Literature and Used in CPA Model

Mode	Task	Time (ms)	Time Used in Model	Reference
Visual	Locate target on screen	1300–3600	Visually locate single target: 1300	Stanton and Baber (2008)
	Recognise familiar words or objects	314–340	Visually locate sequential alphanumeric target: 340	Olsen and Olsen (1990)
	Check if on-screen target is highlighted	600–1200	Check if target is highlighted: 900	Pickering et al. (2007)
	Read page of text on screen, e.g., navigation warning	5000	Read navigation warning: 5000	Average from performing task
	Read number (e.g., temperature) on centre console display	1000–1200	Check temperature display: 1000	Wierwille (1993)
Physical	Move hand from steering wheel to touch screen/remote touch controller, and vice versa	900	Move hand to touch screen: 900	Mourant et al. (1980)
	Press button/target	200	Press touch screen target: 200	Baber and Mellor (2001)
	Move hand between targets	400	Homing on target (movement time during visual search assumed extra): 320 [total 520 with touch target time]	Card et al. (1983)
		505–583		Ackerman and Cianciolo (1999)
		520		Stanton and Baber (2008)
		368–512		Rogers et al. (2005)
	Move pointer to target on screen	1290	1290 includes pressing enter; therefore, positioning time: 1290–570 = 720	Card et al. (1978)
	Press hard enter button	570	Press enter button on remote controller: 570 Press button on centre console: 570	Card et al. (1983)
Auditory	Listen for feedback to confirm correct radio station	3000	Listen for radio station confirmation: 3000	Average from performing task
	Listen for change in audio settings (e.g., bass)	3000	Listen for audio settings confirmation: 3000	Average from performing task
Cognitive	Make simple selection	990	Make selection: 990	Stanton and Baber (2008)

Trade-Off between Context and Objectivity in an Analytic Evaluation

The forward pass calculates the EST and EFT of each operation, moving progressively through the task diagram from left to right (Harrison, 1997). The EST of operation 'X' is determined by the EST of the preceding operation plus its duration. If there is more than one preceding operation that links into operation 'X', then the EST is determined by the latest EFT of the preceding activities:

EST of operation 'X' = EST of preceding operation + Duration of preceding operation

The EFT is the EST of an operation plus its duration time:

EFT of operation 'X' = EST of operation 'X' + Duration of operation 'X'

The backward pass calculates the LST and LFT of each operation, starting from the 'End' node and moving from right to left back through the task diagram. The LST of operation 'X' is determined by the LST of the succeeding activity minus the duration of operation 'X' (Harrison, 1997). If there is more than one succeeding operation that links directly into operation 'X', then the earliest possible LST should be used:

LST of operation 'X' = LST of succeeding operation − Duration of operation 'X'

The LFT of an operation is determined by the sum of the LST and duration of an operation:

LFT of operation 'X' = LST of operation 'X' + Duration of operation 'X'

After calculating the values from the forward and backward passes, the free time, or 'float', is calculated. All paths through the task network, with the exception of the critical path, will have some associated float time. Float time of operation 'X' is the difference between the LST and EST of operation 'X':

Float time of operation 'X' = LST of operation 'X' − EST of operation 'X'

The final stage of CPA involves defining the critical path and calculating total task time. The critical path occurs along the path of operations that has the most minimal float time; in the CPA diagram this is denoted by the solid lines. The durations of all operations on the critical path are summed to produce the total task time.

CPA for IVIS Evaluation

Total task times were calculated for the touch screen and remote controller input devices and are presented in Table 5.6, along with the differences between the two devices for each task.

The CPA method predicted that five tasks would take longer with the touch screen than the remote controller and that two tasks would take longer with the remote

TABLE 5.6
Total Task Times for Secondary Tasks Performed via the Touch Screen and Rotary Controller

	Task Time (ms)		Difference	
Task	Touch Screen	Remote Controller	Remote Controller– Touch Screen	%
Play radio station (909AM)	8460	10770	2310	27.30
Increase bass by 2 steps	11380	11100	−280	−2.46
Increase temperature by 1 degree	4860	4860	0	0.00
Reduce fan speed by 2 steps	5340	6650	1310	24.53
Direct air to face and feet	8880	6080	−2800	−31.53
Direct air to face only	7060	6080	−980	−13.88
Turn on auto climate	3090	3090	0	0.00
Enter destination from system memory	16820	11260	−5560	−33.06
Enter destination from previous entries	16820	11260	−5560	−33.06
Total task time	82710	71150	−11560	−13.98

controller than the touch screen. There was no difference between the two systems in the task times for the 'increase temperature' and 'auto climate' tasks: this was because they were performed via centre console controls rather than the screen-based IVIS, and the task design was identical in both cases. The two air direction tasks were predicted to be shorter with the remote controller than the touch screen. In the remote controller system the user is allowed to select the exact options directly because there are separate options for air to 'face and feet' and 'face only'; however, the touch screen presents three options ('face', 'feet', and 'windscreen'), and the user therefore needs to select multiple options to set air direction to face and feet. The destination entry tasks were also predicted to take longer with the touch screen compared to the remote controller. This is because the touch screen system required users to read a warning about using the navigation warning whilst driving, and this contributed a large amount of time to the task (5000 ms to read the warning, 1300 ms to locate the 'Agree' button, 320 ms homing time to target, 200 ms to touch target: 6820 ms total extra time). Without this extra task segment, the touch screen would have produced a shorter task time prediction for the navigation tasks, compared with the remote controller. Similarly, the time difference in the 'increase bass' task can be attributed to an extra task segment in the touch screen task: with this system, the user has to select the 'Audio/TV' button, then 'Settings', followed by 'Sound', in order to access the 'Bass +' target; however, with the remote controller system, the 'Settings' menu is eliminated, and the user moves directly from the 'Audio' menu to the 'Sound' screen. The time differences between the two IVIS for the air direction, navigation, and increase bass tasks resulted from differences in task design between the two systems. In other words, it is the extra steps involved in the touch screen tasks that were responsible for the observed differences in task times, rather than differences in the nature of the input device. These structural differences

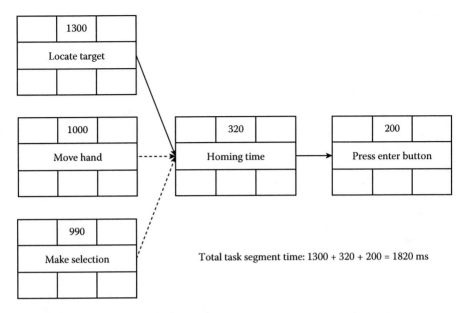

FIGURE 5.5 Excerpt from touch screen CPA diagram to show a single target selection segment.

between tasks, that is, extra task segments, were also identified by the HTA for the same tasks, so the CPA supports the findings of the CPA. However, the CPA also adds another dimension of information to the analysis, and, unlike the HTA, can be used to highlight the effect of input device type on IVIS performance. This is demonstrated by examining the tasks 'play radio station', 'increase temperature', 'reduce fan speed', and 'turn on auto climate'. CPA predicted shorter times for these tasks with the touch screen, compared to the remote controller. When the individual task segments are examined, it appears that the nature of inputs to the touch screen system supports quicker performance because the individual operations have shorter durations, as illustrated in Figure 5.5.

This can be compared to a task segment from a remote controller CPA diagram, showing the same target selection activity, as illustrated in Figure 5.6.

The location options are of equal duration for both IVIS as this operation requires the user to visually locate a target on screen, and the target and screen sizes were approximately equivalent for the two systems so search time would be expected to be the same. The difference in segment time is produced by the second and third operations in the sequence, which involve the user either homing their hand/fingers to the touch-screen target and pressing the target, or manipulating the remote controller to move the cursor to an on-screen target and pressing the enter button on the side of the controller. The touch screen operation times were based on times for moving the hand (320 ms) and pressing a key (200 ms) reported by Stanton and Baber (2008), and there is also some assumed movement of the hand, which occurs in parallel with the visual search operation. Previous studies have reported times of between 368 ms and 583 ms for physical selection of on-screen targets, combining movement

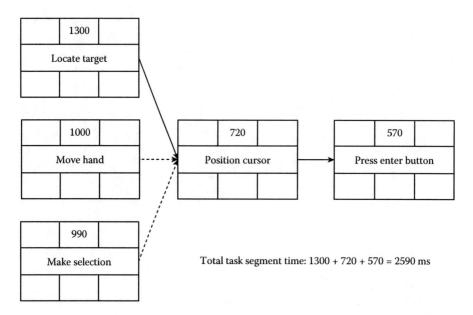

FIGURE 5.6 Excerpt from remote controller CPA diagram to show a single target selection segment.

of the hand and pressing a target (Ackerman and Cianciolo, 1999; Rogers et al., 2005; Stanton and Baber, 2008), which are commensurate with those used in the current study (320 + 200 = 520 ms). Card et al. (1983) reported a time of 570 ms for pressing a pushbutton, and this value was used in the remote controller model for the time to press the enter button on the side of the controller. An assumption was made that pressing a hard enter key located on the side of the remote controller (570 ms) would take longer than touching a target on screen (200 ms) due to the increased resistance from the remote controller button and the reduced ease of access. Card et al. (1978) reported positioning time for a mouse-controlled cursor as 1290 ms, which included target selection via a button press. The movement of the remote controller was very similar to a mouse, and it was assumed that this value provided a good approximation of positioning time for the remote controller. Time to press the enter button (570 ms) was subtracted from total mouse positioning time (1290 ms) to give a value of 720 ms, which was assigned to the positioning of the remote controller in the model. This combination of positioning the cursor and pressing the enter button resulted in longer task-segment times for the remote controller, compared with the touch screen, demonstrating that the nature of the interaction styles of the two devices had an effect on total task times. It is also likely that the GUI layouts had an influence on interaction times, as the size of a target and distance moved between targets influences overall movement time (Fitts, 1954). In this study, it is assumed that the GUI layouts were optimised for each input device, and this would minimise any bias towards a particular GUI layout in terms of the CPA results. Furthermore, average operation timings from HCI literature were used to calculate task times, and these did not account for specific variations in GUI layout between the touch screen and remote controller IVIS. The conclusion that input device type effects interaction times is therefore dependent on the assumption that the GUI layout is optimal for the input device within a particular IVIS.

CPA Utility

CPA enabled quantitative comparisons of task times to be made between the two IVIS, following a structured procedure based on information from the HTA. As this procedure was applied to both systems, it is likely that the relative comparisons, that is, that the remote controller produced consistently longer task times than the touch screen, had high construct validity. This is supported by the idea that indirect input devices that involve some translation between user and on-screen actions are generally slower than direct input devices, which do not involve any translation (Rogers et al., 2005; Stevens et al., 2002). On the other hand, it is not clear if the results represent accurate measures of absolute task times, because they have not been validated against real interactions. There is potential for the CPA method to model absolute task times accurately if a comprehensive and valid database of IVIS operation types could be developed. CPA in its current form also fails to address the issue of the dual task driving environment, as it does not account for breaks in task performance caused by the driver's attention reverting back to the primary driving task. Although stationary IVIS interaction times have been found to correlate well with eyes-off-road time (Green, 1999; Nowakowski and Green, 2001), incorporating the split in visual attention into the model would produce more accurate predictions of IVIS task times in a dual-task environment, that is, IVIS interaction whilst driving.

SYSTEMATIC HUMAN ERROR REDUCTION AND PREDICTION APPROACH (SHERPA)

SHERPA was applied to the two IVIS and operations were classified into one of five types: action, retrieval, checking, information communication, and selection (Stanton, 2006). This classification was based on the analyst's judgement. Within each error type there are a number of error modes, which are shown in Table 5.7.

Each operation in the task HTAs was analysed against the error descriptions in order to identify credible error modes (Baber and Stanton, 1996). When a potential error was identified, the form that the error would take was described according to the analyst's knowledge of the IVIS. The consequences to task performance and recovery potential of the error were then identified: exploring these factors helped in assigning a level of severity to the error (Kirwan, 1992a). Next, the analyst estimated the *probability* of the error occurring during the task and also the *criticality* of the error, using an ordinal scale: low (L), medium (M), high (H). Finally, the analyst proposed remedial strategies to reduce the identified errors. An extract of a SHERPA output for the touch screen IVIS task 'play radio station' is presented in Table 5.8.

SHERPA for IVIS Evaluation

SHERPA was performed on each of the nine tasks for both systems. Tables 5.9 and 5.10 present all identified errors and error modes for the touch screen and remote controller, respectively. The tables also include the probability and criticality ratings for each error, shown in bold. Significant errors were defined as those with either

TABLE 5.7
SHERPA Error Modes and Their Descriptions

Error Mode	Error Description
Action	
A1	Operation too long/short
A2	Operation mistimed
A3	Operation in wrong direction
A4	Operation too much/little
A5	Misalign
A6	Right operation on wrong object
A7	Wrong operation on right object
A8	Operation omitted
A9	Operation incomplete
A10	Wrong operation on wrong object
Information Retrieval	
R1	Information not obtained
R2	Wrong information obtained
R3	Information retrieval incomplete
Checking	
C1	Check omitted
C2	Check incomplete
C3	Right check on wrong object
C4	Wrong check on right object
C5	Check mistimed
C6	Wrong check on wrong object
Information Communication	
I1	Information not communicated
I2	Wrong information communicated
I3	Information communication incomplete
Selection	
S1	Selection omitted
S2	Wrong selection made

high probability or criticality ratings or where both probability and criticality were rated as medium: see Figure 5.7. Significant probability and criticality ratings are highlighted in large/bold in Tables 5.9 and 5.10.

The number of error descriptions—for instance, 'System does not recognize touch'—which were rated as significant was used as a metric by which to compare the two IVIS interfaces. For the touch screen IVIS the SHERPA analysis identified six different error descriptions with significant probability/criticality ratings, compared to seven error descriptions with similarly high probability/criticality ratings identified for the remote controller IVIS. Both systems had the same two errors

TABLE 5.8
Extract of SHERPA Output for Touch Screen IVIS Task 'Play Radio Station'

Task	Error Mode	Error Description	Consequence	Recovery	Probability	Criticality	Remedial Strategy
1 Play radio	N/A	N/A	N/A	N/A	N/A	N/A	N/A
1.1 Open AUDIO/TV menu	N/A	N/A	N/A	N/A	N/A	N/A	N/A
1.1.1 Move hand to touch screen	A8	Driver cannot remove hand from wheel due to high primary task demand	Cannot perform any interaction with touch screen	Immediate, when primary demand allows	M	M	Reduce need for removing hands from wheel – increase number of steering wheel controls, increase automation of secondary tasks
	A9	Driver starts to move hand towards screen but has to replace on wheel due to sudden primary task demand	Cannot perform any interaction with touch screen	Immediate, when primary demand allows	M	H	Reduce need for removing hands from wheel—increase number of steering wheel controls, increase automation of secondary tasks
1.1.2 Prepare to open menu	N/A	N/A	N/A	N/A	N/A	N/A	N/A
1.1.1.1 Make selection	S2	Wrong selection made	Incorrect menu opened	Immediate	L	M	Ensure labels clearly relate to function
1.1.1.2 Locate AUDIO/TV icon	R1	Visual check is not long enough to locate icon	Cannot open desired menu	Immediate, when primary demand allows	M	L	Make icons and labels larger to ensure quick identification
	R2	Incorrect icon is located by mistake	Wrong menu is opened if mistake is not realised	Immediate	L	M	Ensure icons clearly relate to function
1.1.3 Touch AUDIO/TV button	A4	System does not recognise touch	Audio/TV does not open	Immediate	H	L	Increase sensitivity of touch screen
	A6	Touch incorrect button or other part of screen	Incorrect input made or no input made	Immediate	M	M	Increase size of buttons

TABLE 5.9
Errors Identified By SHERPA Analysis for Touch Screen IVIS, Including Probability, Criticality, and Frequency Ratings

Error Mode	Description	Probability	Criticality	Frequency
A2	Consecutive presses are too quick	M	L	3
A4	System does not recognise touch	H	L	24
A4	Press centre console button with too little force	M	L	2
A4	Repeat centre console button press too many times whilst waiting for accurate feedback	L	M	2
A5	User moves hand to wrong area of screen	M	L	17
A6	Touch incorrect button or other part of screen	M	M	24
A6	Touch incorrect button or other part of centre console	H	M	2
A8	Driver cannot remove hand from wheel due to high primary task demand	M	M	9
A8	Driver does not move hand back to steering wheel	L	L	9
A9	Driver starts to move hand towards screen but has to replace on wheel due to sudden primary task demand	M	H	9
A9	Operation incomplete, due to increased demand from primary task	M	M	17
R1	Visual check is not long enough to locate icon	M	L	26
R2	Incorrect icon is located by mistake	L	M	26
C1	Check omitted	L	L	9
C2	Check is not long enough to obtain accurate feedback	L	M	2
S2	Wrong selection made	L	M	26
			Total errors	207

which were of most concern in terms of their probability/criticality ratings. 'Touch incorrect button or other part of centre console' was rated as having a high level of probability and a medium level of criticality for both systems. This is because the location of centre console controls, significantly below the driver's line of sight, means that the driver may have more difficulty locating the controls, compared with targets on screen. If this error occurs, the implications for task performance are critical because no interaction can be performed until the controls are successfully located. The second important error was 'Driver starts to move hand towards screen/ controller ...,' which was rated as being of medium probability and highly critical for both systems. The driver's primary task is to maintain control of the vehicle, and this primary task can often interrupt the interaction with the IVIS, particularly if there is a sudden increase in primary task demand. The implications of this on completing the IVIS task are critical because no interaction can occur until the demand from primary driving has reduced to an acceptable level.

TABLE 5.10
Errors Identified By SHERPA Analysis for Remote Controller IVIS, Including Probability, Criticality, and Frequency Ratings

Error Mode	Description	Probability	Criticality	Frequency
A4	Press button with too little force	L	M	22
A4	Repeat button press too many times whilst waiting for accurate feedback	L	M	2
A5	Pointer misses icon/button/letter/number	H	L	20
A6	Select incorrect icon/button/letter/number	M	M	20
A6	Press down controller instead of enter button located on side of controller	H	L	20
A6	Touch incorrect button or other part of centre console	H	M	2
A8	Driver cannot remove hand from wheel due to high primary task demand	M	M	9
A8	Driver does not move hand back to steering wheel	M	L	9
A9	Driver starts to move hand towards controller but has to replace on wheel due to sudden primary task demand	M	H	9
A9	Driver cannot locate controller after physical search	L	H	7
R1	Visual check is not long enough to locate icon	M	L	22
R2	Incorrect icon is located by mistake	L	M	22
C1	Check omitted	L	L	7
C2	Check is not long enough to obtain accurate feedback	L	M	1
S2	Wrong selection made	L	M	22
			Total errors	194

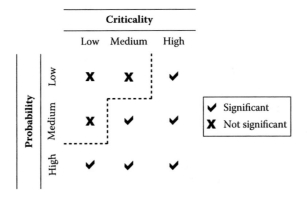

FIGURE 5.7 The threshold defining significant probability/criticality ratings.

SHERPA Utility

Within each system, many of the same errors were identified for each task because the tasks consisted of similar steps. In this study, it is likely that all of the errors identified for both systems would have been identified from an analysis of only one or two representative tasks, which would reduce analysis time considerably. This should be a consideration for future development of error analysis techniques in this context. SHERPA was useful for investigating IVIS interactions in a dual-task environment, that is, performing IVIS tasks at the same time as driving. Instances of incomplete tasks and failure to start tasks were predicted for situations in which the demand from primary driving was high; however, SHERPA provided no way of estimating the severity of these errors or the frequency with which they might occur. Although SHERPA follows a fairly rigid structure for assigning errors, the suggestions for remedial strategies for addressing those errors are likely to differ between analysts (Kirwan, 1992a). SHERPA would benefit from repeated analyses by different personnel on a small sample of representative tasks. A focus group scenario, comprising a mix of ergonomists, designers, and engineers, would also be a useful addition to the method to generate more useful remedial strategies.

HEURISTIC ANALYSIS

The Heuristic Analysis was applied by the analyst, using an adapted IVIS checklist originally developed by Stevens et al. (1999). The checklist was organised into nine sections covering integration of the system into the vehicle, input controls, auditory properties, visual properties of the display screen, visual information presentation, information comprehension, menu facilities, temporal information, and safety-related aspects of information presentation. The evaluation was based on the analyst's experience, gained from four–five hours of interaction with each system, in a stationary vehicle.

Heuristic Analysis for IVIS Evaluation

Tables 5.11 and 5.12 list the issues identified via the Heuristic Analysis for the touch screen and remote controller IVIS. The issues were categorised by the evaluator as positive or negative and further categorised according to the estimated severity of each issue. Negative (major and minor) and positive (major and minor) issues were identified for both IVIS using the heuristic checklist. The number of positive issues identified was the same (6) for both systems. There were slightly more negative issues identified for the remote controller (8), compared to the touch screen (7); however, because the difference was so small and the analysis was purely subjective, it was not possible to use these values to make a valid, quantitative comparison between the two systems.

Heuristic Analysis Utility

The Heuristic Analysis generated qualitative data relating to positive and negative features of each IVIS according to the checklist (Stevens et al., 1999). There are a number of checklists and guidelines for IVIS design (see, for example, Alliance of

TABLE 5.11
Issues Identified through Heuristic Analysis of the Touch Screen IVIS

	Minor	**Major**
Negative	Little use of colour coding Climate menus are cluttered Discomfort from holding arm outstretched when operating touch screen	Glare and reflections on screen Easy to activate wrong control, especially with small buttons, and those which are close together No non-visual location of touch screen controls Small delay in screen response after touch screen interaction for some functions
Positive	Auditory feedback for touch screen button presses Audio volume adjustable for button beeps, parking aid and phone over large range Pop-up screens indicate extra information, e.g., to inform user that phone is disconnected Activation of functions via hard controls is confirmed by a message displayed on screen, e.g., increase/decrease temperature	Text information is easy and quick to read and understand (no long words or cluttered buttons) Easy to go back between menu levels and to return to HOME menu

Automobile Manufacturers, 2006; Bhise et al., 2003; Commission of the European Communities, 2008; Japan Automobile Manufacturers Association, 2004; Stevens et al., 1999); however, no single set of criteria has been accepted as the industry standard. This reflects the difficulty in defining a set of heuristics that is capable of providing a comprehensive checklist for IVIS usability. One of the main problems with the method was the lack of information regarding the frequency with which particular usability problems would occur in everyday usage. A further limitation of the heuristic method is the requirement for a fully developed product or prototype in order to evaluate some aspects of usability. This includes the effect of glare on the IVIS display screen, which cannot be assessed without exposing the IVIS to particular environmental conditions. This is a constraint imposed by the design of many existing checklists for IVIS evaluation, of which the Stevens et al. (1999) checklist is an example; however, it is possible that heuristics could be aimed at an earlier stage in design, eliminating the need for high-fidelity prototypes. For example, Nielsen's Ten Usability Heuristics (2005) encourage a more general approach to usability evaluation, which could be applied in the very earliest stages of product development. Based on these limitations, it is suggested that Heuristic Analysis could be a useful tool for reminding designers about important usability issues (Mendoza et al., 2011; Olson and Moran, 1996), rather than for making direct comparisons between interfaces. The technique has potential for further development by individual automotive manufacturers for making checks on a design to ensure that certain brand- or product-specific targets have been met. The flexibility of Heuristic Analysis means

TABLE 5.12
Issues Identified through Heuristic Analysis of the Remote Controller IVIS

	Minor	Major
Negative	Automatic exit from menu after relatively short time	BACK button located in top, right corner of screen: not quick/easy to access. No hard back button
	Colour coding of menu is not helpful	Relative complexity of navigation menus
	Auditory feedback volume not adjustable for button presses (may not be heard whilst driving)	
	Cluttered appearance of navigation screen (may be unnecessary to display map on screen at all times)	
	Some button labels are unclear, e.g., number input button for navigation	
	Units for inside temperature are not displayed (although it is obvious that numbers refer to temperature)	
Positive	Audio feedback is useful, also indicates an incorrect entry	Screen is well recessed, glass protects screen from glare, screen is located in a 'natural' position for quick glances
	Large text size (compared to other IVIS)	Hard button for HOME MENU (quick access)
	Sensible use of abbreviations (only with functions that require brief glances)	Hard controls are easy to locate non-visually

that specific usability criteria, defined by manufacturers for particular products, could be built in to the checklist.

LAYOUT ANALYSIS

Layout Analysis was performed for a number of IVIS menu screens that were identified by the other analytical methods as having usability issues. This involved grouping similar functions and arranging the layout of these groups according to three factors: frequency, importance, and sequence of use (Stanton et al., 2005). A revised GUI design was created for certain tasks, based on the optimum trade-off between these factors (Stanton and Young, 1998b).

Layout Analysis for IVIS Evaluation

A Layout Analysis for one example menu screen is presented to illustrate the process: see Figure 5.8. The CPA showed that the task time for adjusting fan speed with the remote controller system was reduced when using hard controls, compared with the screen-based controls, suggesting that the design of this menu screen was not optimal. The most significant recommended design change to this menu screen was to reduce the size of the fan speed controls, which had low frequency and importance of use, and to increase the size of the air direction controls, which were used

Trade-Off between Context and Objectivity in an Analytic Evaluation

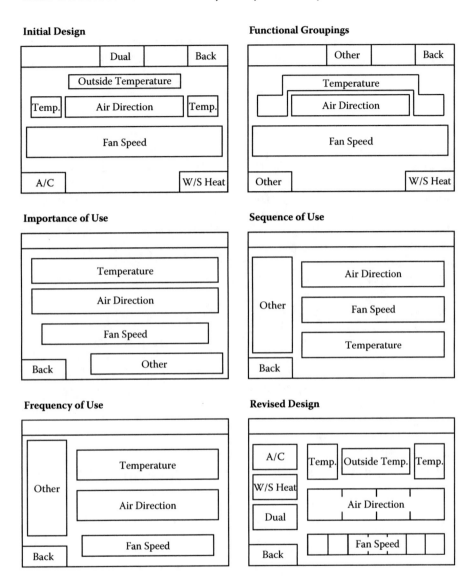

FIGURE 5.8 Layout Analysis for the remote controller climate menu.

more frequently. In order to make a quantitative comparison between the two input types, the number of layout changes made to each system was used as a metric. The two poorest-performing menu screens for each IVIS (including the remote controller climate screen) were identified according to the results of the analytic methods. Layout Analysis was performed on the four menu screens and the number of changes recorded. A change was defined as the movement of a single menu target to a new location according to one or more of the layout rules. The total number of targets that had changed location was calculated and used to compare the two IVIS. There were 11 changes in total to the two touch screen menus which were investigated in this case study, compared with 18 for the two remote controller menus. This could be an indication that in their current forms, the remote controller menu screens would produce a less effective and efficient interaction than the touch screen menus. However, Layout Analysis is highly subjective and in this study was more useful for producing design recommendations rather than direct comparisons of usability.

Layout Analysis Utility

Layout Analysis was included in the method set to provide a technique for specifying design changes based on Ergonomics principles. It is not a useful technique for contrasting different systems as the changes made to a screen according to the rules of layout can be fairly subjective and counting the number of changes to menu target locations is therefore not a valid metric by which to make quantitative comparisons. One use of the technique would be to bridge the gap between evaluation and design: the selection of menus that require redesign is based on the results of the analytic models (CPA and SHERPA), and the redesign is aimed at addressing the issues identified. Layout Analysis would also be useful at very early stages of design, before the prototyping phase, to assist in initial layout decisions (Stanton and Young, 1999a).

GENERAL DISCUSSION

The analytic methods applied in this case study were selected to model the effectiveness of two IVIS. Training, data collection and application times were estimated based on the current study: these will be useful to designers and analysts in future applications of these methods. Training time estimates are presented in Table 5.13, and data collection and analysis times are presented in Table 5.14. The training and application times are similar to those observed in previous studies which have applied these methods (Stanton and Young, 1999a; Baber and Stanton, 1996), with the exception of the application time for Layout Analysis, which in this case was slightly longer than that predicted by Stanton and Young (1999a). This was probably caused by differences in the interfaces tested in the two studies. Dynamic,

TABLE 5.13
Training Time Estimates for the Analytic Methods

Not Much Time	Some Time	Lots of Time
Heuristic Analysis	Layout Analysis	CPA
Familiarisation with checklist	*Learn layout factors*	*Learn rules, calculation method*
<1 hour	*1–2 hours*	*>2 hours*
		SHERPA
		Familiarisation with error codes
		>2 hours
		HTA
		Learn structure and notation
		>2 hours

TABLE 5.14
Data Collection and Application Time Estimates for the Analytic Methods

Not Much Time	Some Time	Lots of Time
Heuristic Analysis	Layout Analysis	HTA
1 hour data collection/	*1–2 hours data collection/*	*2–4 hours data collection/*
1 hour analysis	*1 hour analysis per menu screen*	*6–8 hours data analysis*
		CPA
		2–4 hours data collection/
		8–10 hours data analysis
		SHERPA
		2–4 hours data collection/
		8–10 hours data analysis

screen-based interfaces, analysed in the current study, comprise many different menu layouts in a single system and analysis is therefore likely to be more complex, compared to the static, dashboard mounted controls analysed by Stanton and Young (1999a). In comparison with empirical methods which usually require a sample of users interacting with a prototype product, the time and resource requirements of the analytic methods are significantly lower. This supports their application at an early stage in the product development life cycle. Performance issues identified at this stage can then be further investigated, if necessary, using empirical techniques at a later stage of development when prototypes are more accessible.

Quantitative and qualitative information was extracted from each of the methods in order to make comparisons between the two interaction types under investigation. These data are presented in Table 5.15. Taking the quantitative data in isolation, it could be concluded that HTA and CPA support the use of the remote controller over the touch screen, and the other three measures, SHERPA, Heuristic Analysis, and Layout Analysis, favoured the touch screen over the remote controller. Exploration of the individual methods, however, has shown that it is not sensible to evaluate the IVIS based on this data alone and that some of the methods were unsuitable for making direction comparisons between the touch screen and remote controller interfaces. The findings of this study underline the importance of considering the relevance of outputs on a method-by-method basis (Stanton and Young, 1999a). If the results are used solely to identify which system is superior, then richer information about wider aspects of usability could be lost. Gray and Salzman (1998) warned that the advantages and disadvantages of analytic methods must be understood in order to mitigate against erroneous claims about system usability.

ANALYTIC METHODS FOR IVIS EVALUATION

HTA produced a hierarchical outline of tasks, which described the smallest operations that a user performs when interacting with a particular interface. This analysis showed that the basic task segments for selecting a menu target consisted of the same

TABLE 5.15
Quantitative and Qualitative Comparisons between the Two IVIS

Method	Quantity/ Quality	Touch Screen	Remote Controller	Best Performance?
HTA	Quantity	145 total operations	113 total operations	Remote controller
	Quality	Most touch screen and remote controller tasks have similar structures but the nature of individual operations is different.		
CPA	Quantity	82710 ms total task time	71150 ms total task time	Remote controller
	Quality	Remote controller times are dependent on the speed of movement through menu options and number of options scrolled through before reaching target. When examined at a task segment level, the touch screen is predicted to produce shorter segment time.		
SHERPA	Quantity	6 significant errors	7 significant errors	Touch screen
	Quality	Remedial measures include increasing the sensitivity and allowing better differentiation between targets for the touch screen, increasing precision of the pointer, and moving the enter button for the remote controller.		
Heuristic Analysis	Quantity	7 –ive/6 +ive issues	8 –ive/6 +ive issues	Touch screen
	Quality	Usability issues include glare on the screen and lack of tactile feedback for the touch screen, poor location of the back button, and complexity of menus for the remote controller.		
Layout Analysis	Quantity	11 layout changes across two menu screens	18 layout changes across two menu screens	Touch screen
	Quality	In both devices menu targets with highest importance and frequency of use should be placed in the most accessible place on screen. Sequence of use of targets in IVIS interactions should also be accounted for.		

number of operations for both systems, and highlighted the effect of task structure on interaction strategies with the two IVIS; that is, the touch screen generally required more target presses to complete tasks than the remote controller. Operations identified by the HTA were then fed in to the CPA and assigned duration times in order to calculate predictions of total task times. Like the HTA, CPA also highlighted the differences in task structure between the two IVIS; however, the CPA also showed that although the number of operations in a task segment was consistent, the operation timings assigned to these operations produced differences between the task times of the touch screen and remote controller. Although there was some overlap between the output of these two methods, both are recommended in IVIS evaluation. HTA is a necessary precursor for other methods, including CPA and SHERPA, and is thought to be a useful exercise for familiarising designers and evaluators with task structures. CPA expands the output of HTA by assigning predicted times to the tasks, and the task time metric is useful for comparing IVIS and for making estimates about the effect of IVIS tasks on concurrent tasks, such as driving. SHERPA highlighted a number of potential errors with both systems that would be useful to a designer at the early stages of product development; the remedial strategies devised as part of the analysis would guide any necessary redesign activities in order to reduce errors in the driver-IVIS interaction. There was, however, a quite significant overlap between the errors which were identified for the two systems, which does not support the use of SHERPA as a comparative evaluation tool. SHERPA is based on an objective task description, and the analysis follows a rigid structure that produces quantifiable results; however the assignment of error frequency and severity is dependent on the analyst's judgement. The remedial strategies recommended as part of SHERPA are also an example of qualitative output. Comparison of Heuristic Analysis with the results of SHERPA, in a process of data triangulation (Gray and Salzman, 1998; Wilson, 2006; Hancock and Szalma, 2004; Mackay, 2004), showed that both methods identified some of the same usability issues; however, SHERPA errors tended to relate to individual operations and issues which may prevent these from being performed successfully, whereas the Heuristic Analysis identified more general issues relating to the system and wider environment, for example, glare and reflections obscuring the display screen. There were also instances where the two methods did not agree, and this has also been found in previous studies (for example, Stanton and Stevenage, 1998). For example, glare on the touch screen would lead to a R1 SHERPA error (information not obtained); however, the SHERPA method, which is based on the HTA specification, does not support the analyst in accounting for environmental factors, and therefore this was not identified as a potential error. The issue of false positive error detection has also been found in studies of SHERPA (Baber and Stanton, 1996; Stanton and Stevenage, 1998); this could encourage unnecessary changes to a design. Heuristic Analysis identifies usability issues, and the assumption is that these will lead to poor usability when the IVIS is used by consumers. However, identification of usability issues is not a guarantee of poor performance (Gray and Salzman, 1998). This problem is compounded by the lack of information about frequency of occurrence of issues in this type of analysis. Layout Analysis was only applied to the two worst-performing menu screens in both IVIS; therefore, it is very difficult to make quantitative comparisons between the touch screen and remote controller based on this information alone. The subjectivity of techniques like Layout Analysis, and also Heuristic Analysis and SHERPA, is a disadvantage in situations where quantifiable metrics are needed so that two or more competing systems can be compared. These techniques also suffer from problems associated with the assumption that the analyst always has implicit knowledge of the context-of-use (Blandford and Rugg, 2002): this is often not the case. However, Layout Analysis still adds to the analytic approach by providing a strategy for exploring existing GUI layouts; this is important as the GUI should be optimised with task structure and input device to produce ideal system performance. It also provides designers with a structured method for addressing the types of usability issues identified by SHERPA and Heuristic Analysis.

CONTEXT VERSUS OBJECTIVITY

Usability evaluation should account for the specific context within which systems are used (Harvey et al., 2011d); however, the results showed that not all of the methods addressed this issue. HTA and CPA were developed for application in a single task environment, which means that in this case the effects of driving on IVIS effectiveness were not modelled. Based on this case study, it appeared that the more a method accounts for the broad effects of context, the more subjective it becomes. On the other hand, a narrow and more objective focus produces quantitative models, which enable direct comparisons between systems to be made (Blandford and Rugg, 2002). For example, CPA allows detailed, quantitative, comparable predictions for a very specific aspect of usability; however, the focus on only one aspect of system effectiveness (task times in a single task environment), means that contextual factors are not accounted for (Bevan and Macleod, 1994). Subjective techniques enable a broader approach, which aims to capture the 'whole essence of user-device interaction' (Stanton and Young, 1999a), and these methods therefore account for context to some extent. However, the qualitative nature of the outputs means that these methods do not drill down to a deep level of detail and are therefore more suited to usability checks (e.g., Heuristic Analysis) or design recommendations (e.g., Layout Analysis and SHERPA), rather than direct comparisons (Burns et al., 2005; Butters and Dixon, 1998; Cherri et al., 2004; Jeffries et al., 1991; Nielsen, 1992; Nielsen and Phillips, 1993; Olson and Moran, 1996).

EXTENDING CPA

To address the trade-off between context and objectivity an extension to CPA that allows consideration of the context-of-use is proposed. CPA measures performance via quantitative predictions of task time rather than relying on the assumption that poor performance will follow on from identification of usability issues (Gray and Salzman, 1998; Mendoza et al., 2011; Bevan and Macleod, 1994). Another advantage of CPA is that it takes a taskonomic approach to modelling HMI (Nielsen, 2006), which means that systems are analysed in terms of the activity or task being performed. On the other hand, the heuristic checklist applied in this study took a taxonomic approach because it analysed elements of an interface based on functional, rather than task-based, categories (Stanton and Young, 1999a; Wilson, 2006). Nielsen (2006) argued that both taxonomies and taskonomies are necessary in design; however, in a dual-task driving context, where interaction with secondary tasks is so dependent on the concurrent demand from driving, the activity-based approach (Wilson, 2006) appears to be the most useful for usability evaluation. CPA, in particular, has potential for analysing these dual-task interactions because the driver's interaction with primary driving tasks can be incorporated into the models in parallel to IVIS operations. This technique could be used as a direct measure of the effectiveness of the user–system interaction in a dual task driving environment.

CONCLUSIONS

The aim of the case study presented in this chapter was to explore an analytic approach to IVIS usability modelling to meet a requirement for early-stage, low-resource product evaluation. The methods were selected to model important aspects of HMI performance: task structure, interaction times, error rates, usability issues, and interface design (Gray and Salzman, 1998). The findings of the study have been discussed in terms of IVIS comparisons, utility of the methods, time and resource demands, and potential for further development. HTA was not useful for making relative comparisons between systems; however, it was found to be an essential starting point for CPA and SHERPA, and was also useful for the exploration of task structure. CPA modelled task interaction times as a measure of performance; however, in its current state it does not account for the dual task-driving scenario. There is however, potential to extend the method to address this issue. SHERPA was expected to yield a comprehensive list of potential errors guided by its structured taxonomic approach; however, assessment of error frequency and severity are still largely open to analyst bias. Data triangulation against the results of the Heuristic Analysis also showed that neither method was comprehensive. Heuristic Analysis is not suitable for comparisons between systems; however, there is potential for development as product- or brand-specific guidance. Heuristic Analysis also has the advantages of low training and application times, which supports its use for early identification of potential usability issues. Layout Analysis appears to be useful for bridging the gap between evaluation and design, and has only moderate time and resource demands, which will enable analysts to not only make quick decisions about product performance but also to make recommendations to improve usability. The findings of this exploratory study have highlighted a trade-off between subjectivity and focus on context-of-use. An extension of the CPA modelling method has been suggested to incorporate analysis of context into a quantitative technique so that more useful predictions of IVIS performance can be made. This is explored in Chapters 7 and 8.

6 To Twist or Poke? A Method for Identifying Usability Issues with Direct and Indirect Input Devices for Control of In-Vehicle Information Systems

INTRODUCTION

The focus of Chapter 3 was on the representation of the factors (tasks, users, and system) that influence system performance within a particular context of use. These factors can be represented analytically, by modelling interactions in order to make predictions about performance, and empirically, by representing system components in a simulated or real-world environment and measuring performance with a sample of users. Chapter 5 presented the results of an analytic assessment of IVIS, which was used to compare two existing IVIS and to explore how analytic predictions could be used in evaluation. These predictions are useful at an early stage in product development, when access to prototype systems or samples of users is more restricted, and can give an indication of the potential usability issues with a product or system. Empirical methods are recommended for later stages of the development process, to further investigate the predictions made by analytic methods, using real users and prototypes in order to represent the system with a greater level of fidelity. This chapter presents a case study in which empirical methods were applied to compare two of the most popular IVIS input device types: touch screen and remote controller. The main aim of this case study was to assess how well empirical evaluation methods, selected and described in Chapter 4, could identify usability issues that are specific to these two input types. A set of empirical methods were selected as part of the work presented in Chapter 4, using the flowchart for method selection. These methods are summarised in Table 6.1. As with the analytic methods summary table (Chapter 5), links are made to the KPIs that are measured by the empirical methods.

TABLE 6.1
Empirical Methods and Related KPIs

KPIs	Empirical Methods	Description
3	Longitudinal control	This relates to the speed of the vehicle and includes specific measures of absolute speed, acceleration, speed variation, and following distances to a lead vehicle.
3	Lateral control	This relates to the position of the vehicle on the road and includes measures of lane deviations, steering wheel reversals, and steering wheel angle.
3	Visual behaviour	This is measured by recording the driver's eye movements during driving tasks. Important measures included the amount of time spent looking at the road compared to the amount of time spent looking internally within the vehicle.
2, 5, 7, 8, 10	Secondary (IVIS) task times	Task times reflect the attentional demand of the IVIS and the time the driver spends without their full attention on the primary driving task. It also indicates the effectiveness of a device.
2, 5, 7, 8, 10	Secondary (IVIS) task errors	This indicates the efficiency of task performance with a device and includes measures of error rate and error type.
3	Driving activity load index (DALI)	This method is used to estimate the driver's workload via a subjective questionnaire, which comprises questions about task demands, attention, and stress.
9, 11, 12	System usability scale (SUS)	This method is used to evaluate the driver's subjective satisfaction with the device, via a set of ten rating statements.

The usability of an IVIS is affected by its HMI, which determines how well a driver can input information, receive and understand outputs, and monitor the state of the system (Cellario, 2001; Daimon and Kawashima, 1996; Stanton and Salmon, 2009). The aim of this case study was to evaluate IVIS HMI, particularly the effect of input device type on usability. Two of the most popular IVIS input devices, touch screen and remote controller, were described in Chapter 3. These two IVIS can be distinguished according to the method of input to the system: the touch screen is an example of a direct input device and the remote controller is an indirect input device (Douglas and Mithal, 1997; Rogers et al., 2005).

DIRECT AND INDIRECT IVIS INPUT DEVICES

IVIS input devices can be categorised as direct or indirect (Douglas and Mithal, 1997; Rogers et al., 2005). This describes the relationship between the user's input to a system and the visible actions performed by the system in response. A touch screen creates a direct relationship between what the hands do and what the eyes see (Dul and Weerdmeester, 2001) because the user touches targets directly on screen. When the control input is remote from the visual display, there needs to be some translation between what the hands do and what the eyes see, and this creates an indirect relationship. The characteristics of direct and indirect IVIS input devices were discussed in detail in Chapter 3. In this case study, a rotary dial was used for the remote input to the IVIS. Many automotive manufacturers, including BMW, Audi, and Mercedes-Benz, currently use a variation on a rotary dial for IVIS input. The dial is used to scroll through menu options and is usually pushed down to select a target. In this case study, the two IVIS input devices (touch screen and rotary dial) used an identical GUI and were tested using the same set of tasks, so that any differences observed would be a feature of the input type, rather than of the GUI design or task structure.

EMPIRICAL EVALUATION OF IVIS USABILITY

There have been many empirical studies of driver distraction and its effect on various aspects of driving performance and workload. Many of these studies have used unnatural or 'surrogate' in-vehicle tasks to represent secondary task demand (e.g., Anttila and Luoma, 2005; Carsten et al., 2005; Harbluk et al., 2007; Jamson and Merat, 2005; Lansdown et al., 2004a). Those that have used natural IVIS tasks have tended to focus only on a single task, such as making a phone call (e.g., Drews et al., 2008; Kass et al., 2007; Reed and Green, 1999) or entering a navigation destination (e.g., Baumann et al., 2004; Chiang et al., 2004; Ma and Kaber, 2007; Nowakowski et al., 2000; Oliver and Burnett, 2008; Wang et al., 2010). There have been few empirical usability evaluations of IVIS input devices using a large and diverse set of natural secondary tasks. Rydström et al. (2005) compared one touch screen and two central controller-based IVIS using a set of 10 natural secondary tasks. Their study used the manufacturer-supplied GUIs associated with each of the IVIS. Whilst this would have resulted in high ecological validity, it did not allow direct comparisons to be made between the different input devices because the structure of tasks was different for each system and this, rather than the nature of the input, may have been the cause of any performance differences. In the case study described in this chapter, the same set of tasks and GUI was used for both input devices, which ensured that usability issues could be attributed to the input device, rather than the task structure. In this case study, ecological validity was less important because the main aim was to assess whether or not the empirical methods were capable of highlighting important usability issues, rather than to produce an absolute assessment of IVIS performance.

SELECTION OF TASKS

A set of 17 tasks were selected to represent the four main IVIS function categories: infotainment, comfort, communication, and navigation (see Table 6.2). Although this task set represents some of the same functions that were investigated in the analytic methods case study (Chapter 5), the structure of the tasks were different in this case. Five tasks were selected in each of the first three functional groups. Only two tasks were selected to represent the navigation group, due to the increased time taken to carry out navigation tasks and the limited functionality available in the prototype system used in this case study. Three tasks were repeated to coincide with a roadway event in the driving simulation: this made a total of 20 tasks. Similar to the analytic method case study (Chapter 5), task selection was governed by four

TABLE 6.2
Task Set for the Touch Screen and Rotary Controller IVIS

Task Category	Tasks
Infotainment	Play radio station
	Increase bass
	Adjust balance
	Select portable audio
	Play CD track
Comfort	Increase fan speed
	Increase fan speed[a]
	Turn on auto climate
	Set air direction
	Reduce seat heat
	Turn off climate
Communication	Digit dial
	Call from contacts
	Call from contacts[a]
	Call from calls made list
	Call from calls received list
	Call from calls missed list
Navigation	Enter destination address
	Enter destination address[a]
	Enter destination postcode

[a] Repeated task to coincide with a roadway event.

factors, defined by Nowakowski and Green (2001): need for the task whilst driving, availability of the task in current IVIS, frequency of task performance, and compliance of the task with the 15-second rule. The functions available in existing IVIS were analysed as part of the analytic methods case study, and only those that would be used during the driving task were selected for this empirical case study. Unlike the analytic methods case study, this case study used the same GUI for both input devices in the evaluation. This meant that task structure was identical in both cases, which resulted in more tasks being suitable for valid comparisons. In the analytic methods case study, task selection was limited because of the differences in task structure, in some cases, between the two IVIS under investigation. Many functions are provided by IVIS; however, some would not normally be needed whilst driving, for example, IVIS, Liquid Crystal Display (LCD), and general vehicle settings. Furthermore, some IVIS guidelines advise that certain high-demand tasks, such as navigation entry, are turned off whilst driving because they present a high risk to safety (Commission of the European Communities, 2008; Green et al., 1995). This is, however, a matter of some controversy among end users, who might demand that interaction with tasks should be at their discretion. Consequently, many automotive manufacturers do allow access to functions such as destination entry whilst driving (Llaneras and Singer, 2002), although it is recommended that these functions are accompanied by a warning to drivers regarding the potential distraction risks (Commission of the European Communities, 2008). In this case study, navigation tasks were included in the task set; however, there were no tasks that required users to monitor dynamic information on the screen, for example, watching TV. Frequency of interaction with tasks was estimated based on the authors' experiences from the analytic methods case study. Interaction frequencies for the tasks selected in this study ranged from low (e.g., adjust balance, enter navigation address) to moderate (e.g., select radio station, adjust fan speed). Higher-frequency functions, such as adjust audio volume, tend to be provided via hard, dashboard-mounted controls, rather than as part of a menu-based IVIS (Llaneras and Singer, 2002). These dashboard controls were not investigated in this case study. The tasks were carried out in a pilot study, which showed that three tasks were likely to exceed 15 second ('digit dial', 'enter destination address', and 'enter postcode'): these tasks were included in the set because they were representative of commonly used tasks and would address the issue of access control. Note that the tasks used in this study were different to those used in the analytic methods case study, so the investigations into the task set were carried out in a pilot study before the main empirical work.

Types of Operation

Tasks were also classified according to the types of operations they involved. Three main IVIS operation types were defined for this case study: discrete selection, alphanumeric entry, and level adjustment. Discrete selection operations involve the user selecting a standard menu item in order to open another menu or to select a function at the end of an input sequence. Performance of discrete operations is affected by the number of alternative menu items displayed at one time (Hick's Law; Hick, 1952), the size of the target (Fitts's Law; Fitts, 1954), the visibility of information displayed on the target (Stevens et al., 2002) and its position relative to the previous menu item in the sequence (Card et al., 1983; Fitts, 1954). Alphanumeric entry operations are a type of discrete operation, but specifically involve entering letters or digits. They are differentiated here because they are usually part of long letter/number sequences, for example, in an address or phone number. The layout of alphanumeric targets is particularly important because there is usually a relatively large number to choose from and selection time needs to be minimised. Because of their large number, alphanumeric targets are also usually relatively small, which increases the precision required for successful operation. Level adjustment operations involve the user increasing or decreasing a value, for example, volume or temperature. This can be achieved by continuous movements of a dial or slider or by repeatedly pressing a single target to produce a certain amount of level change.

IVIS have a menu-based structure; therefore, all of the tasks selected here will involve making one or more discrete selections to navigate through this structure. Three of the tasks used in this case study were selected because, in addition to discrete selections, they also involved alphanumeric entry, and four other tasks were selected because they required some form of level adjustment. In this study, the level adjustment tasks involved repeat presses of a single increase/decrease button, rather than continuous movement of a slider or dial. Task selection was limited by the functionality of the prototype GUI used in this study. However, effort was made to select a broad range of tasks, representing all four IVIS function categories and the three operation types of interest. Rotary controller input devices have been found to be better for precision tasks (Rogers et al., 2005), so it was expected that error rate and task time, particularly for tasks involving alphanumeric entry, which requires increased precision, would be lower with the rotary controller. Indirect devices are also suitable for repetitive tasks (Rogers et al., 2005), and it was expected that the rotary controller would also produce a lower error rate and shorter interaction times for tasks involving level adjustment. Rogers et al. (2005) found that direct devices, such as the touch screen, are better for discrete, pointing tasks. It was therefore predicted that the touch screen would yield shorter task times and lower error rates for those tasks that predominantly involved discrete menu selection tasks.

METHOD

Participants

In total, 20 participants (10 female, 10 male) aged between 21 and 33 (mean = 25, SD = 2.8) years took part in the case study. All participants held a valid driving licence and had at least 1 year's driving experience on UK roads (mean = 5, SD = 3.3). Mode annual mileage for the sample was in the range 0–5000 miles. Participants were all right-handed. Participants were recruited via e-mail advertisements, from a sampling frame of civil engineering students and staff at the University of Southampton. They were each paid £20 for participating in the study. The case study was granted ethical approval by the University of Southampton Research Ethics Committee.

Equipment

The University of Southampton's Driving Simulator

The case study was conducted in the University of Southampton's driving simulator. The simulator is a fixed-based system, consisting of a full Jaguar XJ6 right-hand drive vehicle. The vehicle controls are connected to four computers running STISIM Drive™ (System Technology Inc., Hawthorne, CA, USA) software. The road scene was projected onto three 240 cm × 180 cm screens in front of the vehicle, offering a 160° field of view. The rear-view mirror image was projected onto a screen behind the vehicle. Figure 6.1 shows a driver in the simulator: the LCD screen on which the visual IVIS interface was displayed is positioned to the left of the driver and eye-tracking cameras were located in front of the driver.

The Driving Scenario

The driving scenario used in the study simulated a combination of town, city, and countryside driving environments, consisting of dual-carriageway road, with a combination of curved and straight sections, and with-flow and opposite-flow traffic. The

To Twist or Poke?

FIGURE 6.1 Driver in simulator with LCD screen displaying IVIS interface and eye-tracking cameras.

distance from start to finish was 21.9 km, and participants had to drive the full length of the scenario in each condition. The simulator provided auditory feedback to signal when the vehicle strayed over the road edge and to give an indication of vehicle speed via increases/decreases in engine noise. Drivers were also encouraged to maintain suitable driving speeds by having an almost-constant stream of with-flow traffic in the left-hand lane of the dual-carriageway, forcing the driver to maintain an accurate path in the right-hand lane. There was also oncoming traffic in the opposing lanes to discourage the driver from crossing the centreline. If the vehicle was driven too far over the road edge, a crash would be simulated. This feedback encouraged the participants to drive in a natural way. In simulator studies, participants will be aware that poor performance, such as straying out of lane or speeding, poses little real risk to their safety. It was therefore important to provide feedback to demonstrate to drivers that there were negative consequences of poor driving behaviour (Green, 2005).

In-Vehicle Information Systems

The GUI used in this study was displayed on a 7-inch LCD screen, mounted on the dashboard to the left of the driver (towards the centre of the vehicle). The LCD screen was connected to a laptop in the rear of the vehicle, from which the experimenter could also control the GUI. The LCD enabled touch input. In the rotary controller condition, a rotary input device was mounted just in front of the gear lever in the car's centre console. The controller moved clockwise and anticlockwise to scroll through the on-screen options. It could also be pressed down to select options; however, it did not move forwards/backwards/right/left, and therefore lacked the full functionality of most existing rotary systems, such as the BMW iDrive, Mercedes Command, and Audi Multi Media Interface (MMI). The two input devices evaluated

FIGURE 6.2 Touch screen IVIS input device used in empirical methods case study.

in this study are shown in Figures 6.2 and 6.3. Both IVIS used the same GUI: screen shots of the main menu and climate menu screens are presented in Figures 6.4 and 6.5, respectively. Auditory feedback was not provided for either system.

Eye Tracking

The simulator was equipped with an eye-tracking system (FaceLab™, version 4.6; Seeing Machines, Canberra, Australia), which measured participants' visual behaviour, including time spent looking at the road scene and the LCD display. This system consisted of two cameras and an infrared reflector pod mounted on the dashboard in front of the driver.

User Questionnaires

Each participant was provided with paper copies of a participant information sheet, a demographic questionnaire, a consent form, and SUS questionnaire (Bangor et al., 2008; Brooke, 1996). SUS consists of a 5-point scale, against which participants rated their agreement with 10 statements relating to the usability of a system (see Chapter 4 for a description of the SUS method). An overall score for system usability between 0 and 100 was calculated for each IVIS.

To Twist or Poke?

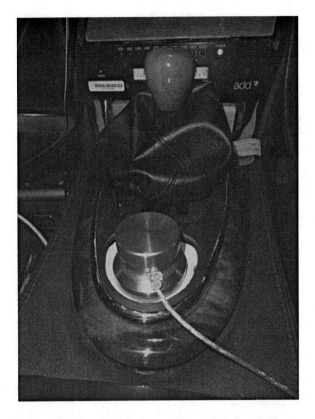

FIGURE 6.3 Rotary controller input device used in empirical methods case study.

FIGURE 6.4 Screen shot of the home menu screen from the prototype IVIS GUI (image reproduced courtesy of Jaguar Cars).

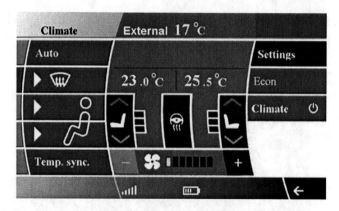

FIGURE 6.5 Screen shot of the climate menu screen from the prototype IVIS GUI (image reproduced courtesy of Jaguar Cars).

Procedure

Pilot studies were conducted in order to refine aspects of the study design, including the length and complexity of the driving scenario, number of tasks, and method of task presentation. In the main case study, participants were first briefed about the experiment and then asked to complete a consent form and a questionnaire to gather demographic and driving experience information. Participants were allowed to adjust the seat and mirror position before the test started. Each participant was then given a 10-minute practice drive in the simulator, during which various vehicle controls and features of the road scenario were explained. Next, participants drove through a simulated driving scenario, lasting approximately 25 minutes. In this control condition, they did not perform any secondary tasks via an IVIS. In the next phase of the experiment, participants completed the two IVIS conditions. Before each condition, participants were given 5 minutes to practice with the IVIS. In each experimental condition, participants drove through the same driving scenario as in the control condition, whilst performing the 20 secondary tasks via each IVIS. A repeated measures design was used, and the order of presentation of IVIS conditions was counterbalanced across participants to eliminate learning and practice effects. After each IVIS condition, participants completed the SUS questionnaire.

Secondary In-Vehicle Tasks

In each experimental condition, participants were instructed to complete 20 tasks whilst driving (see Table 6.2). This set of tasks was the same for the touch screen and rotary controller conditions. Instructions to complete each task were read out to participants by the experimenter, who was seated in the rear of the vehicle. Each task was read out approximately 20 seconds after the participant had completed the previous task. The order of task presentation was randomised for each participant to minimise practice effects. In each condition, three events were triggered by the experimenter to coincide with certain tasks (marked with an asterisk in Table 6.1). These tasks were representative of low (increase fan speed), medium (call from contact list), and high (enter destination address) levels of relative complexity. Levels of complexity were assigned based on analysis carried out via CPA, which was used to explore the number of menu levels, number of operations, and operation types for the two IVIS. The three events were always presented in the same sequence as follows:

1. Man walks out into the road in front of the driver's vehicle, crossing from right to left.
2. Woman walks out into the road in front of the driver's vehicle, crossing from right to left.
3. Dog walks out into the road in front of the driver's vehicle, crossing from left to right.

Participants were not informed about the pedestrian events before any of the trials, or that these events would always coincide with particular secondary tasks. The pedestrian events were triggered by the experimenter. The triggering of events could not be seen by participants; this ensured that they would not be able to anticipate a collision and change their behaviour in response.

DATA COLLECTION AND ANALYSIS

The study employed a repeated measures design. IVIS condition was a within-subjects factor, consisting of three levels: control (no IVIS), touch screen, rotary controller. Primary driving performance data were recorded by the simulation software. This included mean speed and number of centreline crossings. Subjective ratings of system usability were recorded using the SUS questionnaire. A key logger was used to record the specific target and time (in ms) every time the user touched the LCD screen or selected a target using the rotary controller. These data were logged to a file and were used to calculate total task times for the touch screen and rotary controller. Visual behaviour data were recorded by the eye-tracking equipment. The FaceLab cameras tracked the position of each user's head and gaze. Gaze is tracked by reflecting infrared light, which is emitted from a pod located between the cameras in front of the participant, off the participant's eyes, into the cameras. The FaceLab system uses the infrared 'glints' from each eye to derive the participant's gaze vector. To determine the eye's fixation point, the system calculates the point in space where the eye gaze vector intersects with an object in the world. These objects must be defined prior to testing and consist of a set of 3D objects that are used to model the 3D environment observed by the participant. A world model was set up for the simulated driving environment in this study and consisted of objects to represent the three simulator projector screens on which the road view was displayed, the LCD screen on which the IVIS GUI was presented, the instrument cluster, and the rear-view mirror. During each test, the FaceLab system logged the participant's eye fixation point, in terms of one of the predefined world objects, at 30-ms intervals. Visual attention was calculated as the amount of time the eye fixated on each of the world objects, and this was expressed as a percentage of total test time. Head position and gaze tracking was calibrated for each individual participant to ensure high levels of tracking accuracy; however, noise in the data is inevitable due to the inherent instability of the eye (Duchowski, 2007). Performance of the eye-tracking equipment was affected by certain facial features and was less accurate for certain participants, particularly glasses or contact-lens wearers. On average, 4% of fixation points were not tracked to a world object and were logged as noise; however, in four cases, the eye-tracking data contained high levels of noise (in excess of 5% of total fixations were logged as noise) and the visual behaviour data for these four participants were removed prior to statistical analysis.

Primary driving performance and visual behaviour metrics were compared across the conditions using a Friedman's Analysis of Variance (ANOVA), for multiple related samples. The data for all measures were tested for normality and found to be nonnormally distributed; therefore, nonparametric statistical tests were applied. Post hoc tests (Wilcoxon tests for two related samples) were also applied, with a Bonferroni adjustment for multiple comparisons. Effect sizes (r) are also reported in accordance with American Psychological Association guidelines (Wilkinson, 1999). Outliers are shown as a point for values plus/minus 1.5 times the interquartile range (IQR) from the top/bottom whiskers and as an asterisk for values plus/minus three times the IQR from the top/bottom whiskers.

The age of participants in this study ranged from 21 to 33 years. There is some evidence to suggest that drivers aged 25 and under exhibit different driving performance, visual behaviour, and crash risk, under dual task conditions, compared with drivers over 25 (Liu, 2000; Reimer et al., 2011; Ryan et al., 1998). To examine whether or not this was a factor in the current study, each set of results was split by age into two groups: 21–25 year olds (n = 13, mean age = 23, SD age = 1.2, mean experience = 4 years, SD experience = 2.1, mode mileage = 0–5000 miles), 26–33 year olds (n = 7, mean age = 28, SD age = 2.4, mean experience = 8 years, SD experience = 3.9, mode mileage = 5001–10000). The two groups were then compared using Mann Whitney tests for two independent samples. No significant differences were found between the age groups on any of the usability measures reported here. Although it is widely accepted that there are age-related differences in driving performance and distraction caused by interaction with IVIS tasks, these differences may only be significant at the more extreme ends of the scale. For example, Shinar (2008) observed a decline in driving performance with a concurrent mobile phone task only with older adults aged 60–71 years. They found little difference in the performance of two younger age groups, aged 18–22 and 30–33 years. Horberry et al. (2006) also found few age-related performance differences in dual task conditions, particularly for drivers under 60. These findings support the results of the age comparisons in this study, and results are therefore reported across the entire age range, 21–33 years.

RESULTS AND DISCUSSION

Primary Driving Performance

Previous studies have shown that, when drivers interact with secondary in-vehicle tasks, their workload increases and this can often lead to distraction (Dingus et al., 2006; Lansdown et al., 2004a; Lees and Lee, 2007; Jamson and Merat, 2005; Wang et al., 2010). In this case study, driving performance whilst interacting with secondary tasks was compared with a control condition of driving without task interaction. As expected, both IVIS produced significantly worse levels of driving performance, compared with the control condition. This was reflected in measures of mean speed, speed variance, and number of centreline crossings.

Longitudinal Control

The driver has immediate control over their speed and, consequently, this is one of the most significant factors in measuring driver distraction (Bullinger and Dangelmaier, 2003; Collet et al., 2010a; Fuller, 2005). Speed was expected to be lower in the IVIS conditions, with the most distracting/demanding interface causing the largest reduction in speed. Reduction in speed as a result of increased workload has been observed in previous studies of driver distraction (e.g., Green et al., 1993; Jamson and Merat, 2005; Johansson et al., 2004; Lansdown et al., 2004a; Tsimonhi et al., 2004; Young et al., 2003). It is thought that most drivers employ this strategy to reduce primary task workload in order to cope with the demand from the interaction with secondary tasks. Drivers were told to drive at 40 mph consistently throughout each run, and there were 40 mph speed limit signs displayed at regular intervals in the driving scenario. Drivers recorded the highest mean speed in the control condition and the lowest in the rotary controller condition. A box plot comparing speeds across the three conditions is shown in Figure 6.6.

There was a significant effect of condition on mean speed ($\chi^2(2) = 14.70$, $p < 0.001$). The mean speed in the rotary controller condition was significantly lower than in the control condition ($z = -3.21$, $p = 0.001$, $r = -0.51$); however, there was no significant difference between the touch screen and control conditions ($z = -1.33$, $p = 0.96$). Comparisons between the two IVIS showed that mean speed in the rotary controller condition was significantly lower than in the touch screen condition ($z = -2.50$,

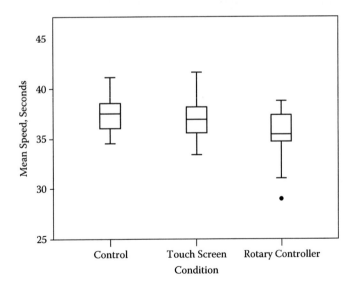

FIGURE 6.6 Box plot of mean speed.

$p < 0.05$, $r = -0.40$). Standard error speed was highest in the rotary controller condition, indicating wide variation in speed between users in the sample when interacting with this device. Speed was more consistent between users with the touch screen.

Lateral Control

Lateral control was measured as the mean number of centreline crossings during each condition. A centreline crossing was recorded every time the wheels of the driver's vehicle made contact with the other side of the roadway. Maintaining trajectory is one of the main driving tasks (Fuller, 2005) and demands high visual attention in particular. If this attention is diverted to secondary tasks, then performance will consequently suffer (Collet et al., 2010a). A box plot comparing the number of centreline crossings across the three conditions is shown in Figure 6.7.

The highest rate of centreline crossings occurred in the rotary controller, followed by the touch screen and finally the control condition. The results showed a significant effect of condition on mean centreline crossings ($\chi^2(2) = 17.22$, $p < 0.001$). Compared with the control condition, there was a significantly higher mean number of centreline crossings by drivers in the touch screen condition ($z = -3.33$, $p < 0.001$, $r = -0.53$) and the rotary controller condition ($z = -3.44$, $p < 0.001$, $r = -0.54$). These results are consistent with findings from previous distraction studies (e.g., Jamson and Merat, 2005; Lansdown et al., 2004a). Of the two IVIS, the rotary controller condition produced the highest rate of centreline crossings ($z = -2.27$, $p < 0.05$, $r = -0.36$). As with other driving metrics, this degradation in lane-keeping performance is thought to be a consequence of reduced attention to the primary driving task. In contrast to the results of this study, Wang et al. (2010) did not detect any significant differences in longitudinal and lateral driving performance between the three IVIS that they tested. They attributed their result to the low level of demand induced by the secondary tasks in their study, which involved users entering a maximum of six

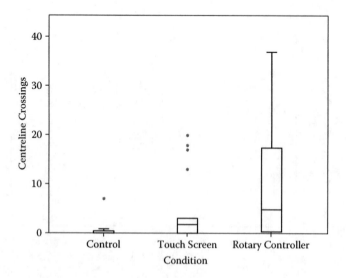

FIGURE 6.7 Box plot of centreline crossings.

characters for navigation entry. In the present study, the navigation and communication tasks consisted of the user entering longer alphanumeric combinations, resulting in a greater duration of secondary task demand. The frequency of task presentation in the case study presented here is also likely to have increased workload, compared with the study by Wang et al. (2010). This study, as with other simulator-based experiments, was designed to compress the experience of secondary task interaction (Stanton et al., 1997), so that the magnitude of effect would be high, allowing usability issues to be identified more easily. In reality, drivers would never interact with such a high frequency of secondary tasks, in such a short period of time. This obviously will have affected the ecological validity of the study presented here; however, this study focussed on how effectively the methods could compare different systems and highlight usability issues and therefore the validity of the testing environment was not a particularly significant factor (deWinter et al., 2009).

VISUAL BEHAVIOUR

The visual mode is the main mode of information presentation from system to human during primary driving (Brook-Carter et al., 2009; International Organization for Standardization, 2002; Sivak, 1996; Victor et al., 2009; Wierwille, 1993). Drivers need to maintain a high level of visual attention to the forward road scene; however, they must time-share this attention with additional objects and events in the visual periphery, such as information displayed on an IVIS (Brook-Carter et al., 2009; Dukic et al., 2005; Pettitt et al., 2005; Pickering et al., 2007; Victor et al., 2009; Wang et al., 2010). The visual demand of secondary IVIS tasks will affect the level of interference with primary tasks and, consequently, the driver's performance (Dukic et al., 2005; Wang et al., 2010). The visual behaviour of participants was monitored in each condition. For all participants, the majority of time during each trial was spent looking at either the forward road scene, which was measured as visual fixations on the left, right, and front projector displays, or the LCD, situated within the vehicle, on which the GUI was displayed. For each condition, time spent looking at the road scene and LCD was measured and then calculated as a percentage of total trial time. Note that percentages do not sum to one hundred as the participants also spent some time during the tests fixating on the instrument cluster and the rear-view mirror. The visual attention to the road scene for the three conditions is shown in the box plot in Figure 6.8.

There were significant differences in visual attention to the road scene between the three conditions ($\chi^2(2) = 32.00$, $p < 0.001$). As expected, drivers spent a significantly higher proportion of time looking at the road scene in the control condition, compared with the two IVIS conditions (both IVIS-control comparisons: $z = -3.52$, $p < 0.001$, $r = -0.62$). The rotary controller condition also produced significantly less visual attention to the road scene than the touch screen ($z = -3.52$, $p < 0.001$, $r = -0.62$).

The LCD data showed that visual attention to the LCD was highest in the rotary controller IVIS condition. Visual attention to the LCD across the three conditions is shown in the box plot in Figure 6.9.

There was a significant effect of condition on visual attention to the LCD ($\chi^2(2) = 30.13$, $p < 0.001$). Visual attention to the LCD in the two IVIS conditions was

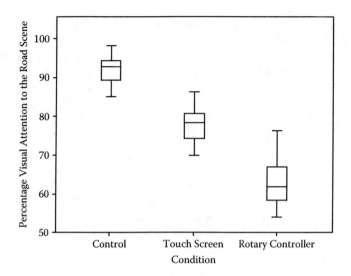

FIGURE 6.8 Box plot of percentage visual attention to the road scene.

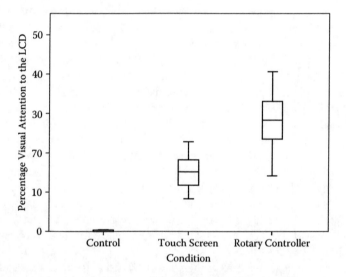

FIGURE 6.9 Box plot of percentage visual attention to the LCD.

significantly higher than in the control condition (both IVIS-control comparisons: $z = -3.52$, $p < 0.001$, $r = -0.62$). In the rotary controller condition, visual attention to the LCD was significantly higher than in the touch screen condition ($z = -3.46$, $p < 0.001$, $r = -0.61$). The rotary controller IVIS produced the worst performance in terms of visual distraction, with highest attention to the LCD and lowest to the roadway. Wang et al. (2010) also identified their scroll-wheel input device as having the worst level of performance according to visual behaviour measures applied in their study. They identified the largest difference in visual behaviour between simulator and real driving environments for the touch screen IVIS. This result was attributed to the effects of glare in the real road driving condition, which increased visual demand to the IVIS. Glare would be likely to affect both interfaces used in the current study, under real road driving conditions, as they used the same LCD screen. In reality, the design of visual IVIS components can reduce the effects of glare, although this will be most difficult for the touch screen, because its LCD cannot be set back and shrouded from sunlight.

SECONDARY TASK PERFORMANCE

Secondary task performance measures reflect the effectiveness and efficiency of the interaction with an IVIS (Sonderegger and Sauer, 2009). These measures were taken during the driving task to evaluate how well the input devices supported the driver–IVIS interaction when drivers were operating in a dual-task environment.

Secondary Task Times

Secondary task times give an indication of the time that a driver spends without their full attention on the road scene. The more time a driver spends interacting with an IVIS, the higher the risk to safe driving (Green, 1999; Wang et al., 2010). High task times also indicate low levels of IVIS effectiveness and efficiency. In each task, performance time was measured from when the driver selected the first IVIS option to when they selected the last option to complete the task. The task times reported in this case study represent error-free tasks that were performed simultaneously with the primary driving task. When users made incorrect operations as part of a task, that is, errors, the total task time was increased because of the additional time taken to make the initial error and then for the operations required to correct the error. As the errors made were not consistent across users, tasks that contained errors could not be used in the mean calculations and were therefore removed from the dataset. Tasks were being performed whilst driving and were therefore often interrupted, so drivers could attend to the primary task. Although the driving scenario was designed to be as consistent as possible, it was impossible to control for the amount of disruption across different tasks and different users. These task times should therefore be interpreted with some caution. The individual times are likely to be significantly longer than static task times, and consistency between tasks, in terms of interruptions, is likely to be low; however, the magnitude of difference between mean task times and across the two IVIS is likely to be accurately represented by these results. For example, for both the touch screen and rotary controller, the results show that relatively simple tasks, such as increase bass and select CD, took considerably less time than more complex tasks, such as enter destination address and digit dial. Table 6.2 shows the mean and standard deviation task times for the 20 tasks performed using the touch screen and rotary controller IVIS. Where a participant made an error in a particular task with just one of the IVIS, the task time data for both IVIS for that participant and task were removed. This was to ensure equal sample sizes across the IVIS conditions. This resulted in some task samples of less than 12, which

TABLE 6.3
Mean and SD Task Times for Error-Free Performance with Touch Screen and Rotary Controller IVIS

	Touch Screen		Rotary Controller					
	Mean (s)	SD (s)	Mean (s)	SD (s)	N	z	p	r
Play radio station	30.70	—	12.42	—	1	—	—	—
Increase bass	10.34	4.29	26.18	12.08	14	−3.23	<.001	−.61
Adjust balance	18.83	17.3	29.42	13.92	11	−1.87	<.05	−.40
Select portable audio	9.91	2.98	29.81	17.09	16	−3.52	<.001	−.62
Play CD track	5.07	1.05	18.38	9.19	15	−3.41	<.001	−.62
Increase fan speed	4.96	2.39	13.64	6.79	17	−3.62	<.001	−.62
Increase fan speed[a]	12.60	11.43	26.91	11.41	15	−2.39	<.05	−.49
Set air direction	14.29	4.89	45.16	17.99	14	−3.30	<.001	−.62
Turn on auto climate	4.52	2.68	8.43	6.18	16	−2.12	<.05	−.37
Reduce seat heat	8.46	5.19	22.63	18.93	13	−3.11	<.001	−.61
Turn off climate	4.50	2.51	12.72	4.18	13	−3.11	<.001	−.61
Digit dial	21.12	1.15	101.18	86.54	3	—	—	—
Call from contacts	11.43	4.50	31.93	20.98	5	—	—	—
Call from contacts[a]	29.28	10.64	32.21	9.84	10	—	—	—
Call from calls made list	8.80	2.67	22.28	8.68	14	−3.30	<.001	−.62
Call from calls received list	10.54	3.59	25.95	19.13	20	−3.81	<.001	−.60
Call from calls missed list	9.19	2.82	23.69	8.67	16	−3.52	<.001	−.62
Enter destination address	27.34	7.66	71.37	19.55	7	—	—	—
Enter destination address[a]	46.09	12.04	107.63	31.92	9	—	—	—
Enter destination postcode	21.71	2.83	85.58	40.42	8	—	—	—

were considered too small for meaningful statistical analysis (Nielsen, 1993; Stevens et al., 2002). Only tasks with samples of 12 or more were analysed (using Wilcoxon tests for two related samples), and these results are also reported in Table 6.3.

The touch screen produced consistently shorter interaction times than the rotary controller. Contrary to the predictions made regarding the suitability of the different input devices to the different operation types, the rotary controller did not produce shorter interaction times for tasks that involved greater precision or repetitive operations. These results indicated that this method of input, that is, turning the dial to highlight an option and pressing down on the dial to select the option, took more time than touching an option on the touch screen, irrespective of task type. In all but three tasks, the standard deviation task time was also larger for the rotary controller, indicating greater variability between users, compared with the touch screen. These findings are supported, in part, by the results of a study by Rogers et al. (2005), which compared task times across a touch screen and a rotary controller. This showed that with younger users task times were shorter for the touch screen for most task types that were assessed, including level adjustments and discrete selections. The picture was less clear for older users and for tasks that involved repetitive operations. As expected in the case study presented here, when a task coincided with an event in the road, task times were longer compared with the same task without the event. Reed-Jones et al. (2008) also reported an increase in secondary task times from a hazard-free driving scenario to a hazardous one, in which other road users entered the driver's projected path. In the current study, when a pedestrian was triggered to cross in front of the vehicle, participants either collided with the pedestrian or avoided it. Both outcomes had a negative effect on secondary task performance, which is reflected in the increased mean task times for the task/event combinations. For example, one participant performed the address entry task (without a concurrent event) in 27.5 s, which is close to the mean time for this task. In the address entry/pedestrian event combination task, the same participant collided with the pedestrian and was forced to interrupt the task to attend to the collision. This increased their task time to 63.3 s. Closer examination of the task revealed that there was an interruption of 31.2 s between two task steps: open destination entry menu, open address entry menu. At this point, the participant was attending to the collision, rather than the task, and this contributed to a much longer task time. Another participant recorded times of 26.7 s and 36.8 s for the address entry task and task/event combination, respectively. This participant managed to avoid colliding with the pedestrian; however, there was still an obvious interruption in the task (19.9 s), during which the participant was attending to the road in order to avoid the pedestrian. These examples support the conclusion that drivers were unable to successfully divide their attention between the primary and secondary tasks when primary demand was increased to a level above normal driving, that is, by a roadway event. In tasks that did not coincide with an event, overall time was shorter and times between consecutive task steps were more consistent, indicating that the driver was able to divide their attention more effectively.

Secondary Task Errors

One of the requirements for an effective and efficient IVIS is a low error rate (see Chapter 2). Making an error means that the intended IVIS function will not operate correctly and will often require the user to identify the cause of the error and perform corrective operations (Card et al., 1983; Nielsen, 1993). This increases the number of inputs into the system and the level of attentional demand required by the secondary task. Errors also frustrate users, leading to low levels of satisfaction (Jordan, 1998a). Error rates for each task are shown in Table 6.4.

Errors are reported per task step: a task step was defined as each new selection of a different menu target. Longer tasks would be expected to produce more errors because there are more steps to successfully carry out; however, this would not reflect the actual difficulty of the task. Some sample sizes were smaller than 20 because some participants did not have time to complete all the rotary controller tasks successfully. Based on previous findings, which support the suitability of indirect devices for precision and repetitive operations (Rogers et al., 2005), it was predicted that the rotary controller would produce fewer errors than the touch screen

TABLE 6.4
Mean and SD Errors per Task Step for the Touch Screen and Rotary Controller IVIS

	Touch Screen Errors per Task Step		Rotary Controller Errors per Task Step				
	Mean	SD	Mean	SD	N	z	p
Play radio station	0.19	2.30	0.17	0.24	18	−0.33	ns
Increase bass	0.05	0.13	0.09	0.15	19	−1.41	ns
Adjust balance	0.03	0.06	0.03	0.07	17	0.00	ns
Portable audio	0.00	0.00	0.04	0.18	17	−1.00	ns
Play CD track	0.00	0.00	0.04	0.11	17	−1.41	ns
Increase fan speed	0.00	0.00	0.06	0.13	18	−1.73	ns
Increase fan speed[a]	0.10	0.24	0.07	0.14	20	−0.71	ns
Set air direction	0.03	0.09	0.04	0.07	18	−1.34	ns
Activate auto climate	0.03	0.11	0.06	0.16	18	−0.58	ns
Reduce seat heat	0.02	0.07	0.04	0.11	14	−0.58	ns
Turn off climate	0.10	0.26	0.06	0.17	17	0.00	ns
Digit dial	0.09	0.14	0.07	0.12	18	−0.51	ns
Call from contacts	0.07	0.10	0.09	0.10	18	−0.58	ns
Call from contacts[a]	0.04	0.08	0.11	0.18	19	−1.31	ns
Call from calls made list	0.01	0.06	0.08	0.31	16	−1.48	ns
Call from calls received list	0.01	0.06	0.01	0.06	20	0.00	ns
Call from calls missed list	0.01	0.06	0.01	0.06	18	0.00	ns
Enter destination address	0.02	0.04	0.04	0.04	17	−1.27	ns
Enter destination address[a]	0.03	0.04	0.04	0.04	20	−1.07	ns
Enter destination postcode	0.02	0.04	0.03	0.05	19	−0.51	ns

for alphanumeric tasks, including enter navigation destination and digit dial, and level adjustment tasks, including balance, bass, seat heat, and fan speed. The results of the case study presented in this chapter showed that the rotary controller produced a lower per-task error rate for the 'play radio station', 'increase fan speed' (with event), 'turn off climate', and 'digit dial' tasks. The fan speed and digit dial tasks involved high precision and/or repetitive operations, and this result supports the prediction. However, the other tasks that were predicted to yield better results with the rotary controller actually produced a higher or equal rate of errors with this device. Error rates were compared using Wilcoxon tests, which showed that there were no significant differences between the touch screen and rotary controller.

SUBJECTIVE MEASURES

System Usability Scale (SUS)
The SUS consisted of 10 statements about different aspects of product usability, against which users rated their agreement (Bangor et al., 2008; Brooke, 1996). A

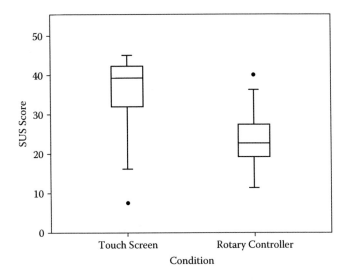

FIGURE 6.10 Box plot of System Usability Scale (SUS) scores.

single usability score for each IVIS was calculated from these ratings. A box plot comparing the scores is shown in Figure 6.10.

The SUS score for the rotary controller was significantly lower than the score for the touch screen ($z = -3.31$, $p < 0.001$, $r = -0.52$), and there was the least variation in this value, indicating a consensus of poor opinion among participants (rotary controller: mean SUS score = 46.88, SD = 14.32; touch screen: mean SUS score = 71.00, SD = 14.32). This result is commensurate with the primary and secondary task performance measures, which showed the touch screen to have better performance and usability than the rotary controller. This indicates that the participants were able to use the SUS to successfully report the trend in the results of the objective usability measures, supporting the use of both types of measures as part of the evaluation framework (Sonderegger and Sauer, 2009).

Driving Activity Load Index (DALI)

DALI was used to measure users' perceptions of their workload whilst driving and performing secondary tasks. The questionnaire assessed seven different factors of workload: global attention demand, visual demand, stress, interference, auditory demand, tactile demand, and temporal demand (Pauzié, 2008). In this study, the first four factors were of most interest and the mean ratings for these factors are shown in Table 6.5.

The results show a trend for each workload factor, with the control condition producing lowest levels of subjective workload (where applicable), followed by the touch screen and finally the rotary controller. This trend supports the results of the performance measures as higher workload would be expected to produce worse primary and secondary task performance. Box plots comparing the ratings are shown in Figures 6.11 (global attention demand), Figure 6.12 (visual demand), Figure 6.13 (stress), and Figure 6.14 (interference).

TABLE 6.5
Mean and Standard Deviation Scores for Four Parameters of the DALI Evaluation

	Control		Touch Screen		Rotary Controller		
	Mean	SD	Mean	SD	Mean	SD	N
Global attention demand	2.20	1.01	3.50	0.83	4.05	1.10	20
Visual demand	2.70	1.17	4.05	8.26	4.35	0.81	20
Stress	1.45	1.15	3.15	1.81	3.95	1.10	20
Interference	—	—	3.70	0.73	4.10	1.25	20

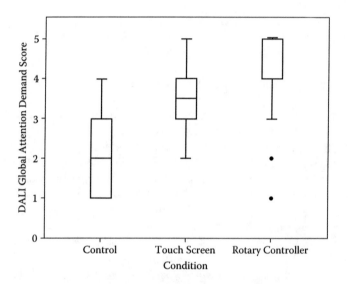

FIGURE 6.11 Box plot of DALI Global Attention Demand ratings.

There was a significant effect of condition on ratings of global attention demand ($\chi^2(2) = 26.11, p < .001$). Compared with the control condition, workload ratings were significantly higher in the touch screen condition ($z = -3.57, p < .001, r = -.43$) and the rotary controller condition ($z = -3.70, p < .001, r = -.48$). Global attention demand was rated significantly higher in the rotary controller condition, compared to the touch screen condition ($z = -2.23, p < .05, r = -.29$).

There was a significant effect of condition on ratings of visual demand ($\chi^2(2) = 27.03, p < .001$). The touch screen produced significantly higher visual demand ratings than the control condition ($z = -3.72, p < .001, r = -.48$). The rotary controller also produced significantly higher ratings, compared with the control ($z = -3.67, p < .001, r = -.47$). There was no significant difference in the visual demand ratings between the touch screen and rotary controller ($z = -1.26, p = .143$). There was a significant effect of condition on ratings of stress ($\chi^2(2) = 27.70, p < .001$). Compared

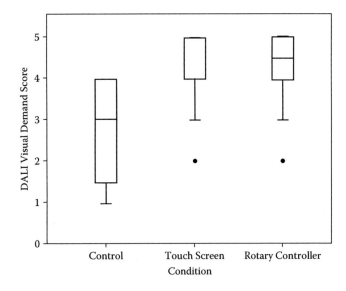

FIGURE 6.12 Box plot of DALI Visual Demand ratings.

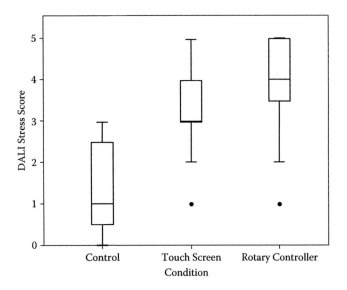

FIGURE 6.13 Box plot of DALI Stress ratings.

with the control condition, stress ratings were significantly higher for the touch screen ($z = -3.47$, $p < .001$, $r = -.45$) and the rotary controller ($z = -3.96$, $p < .001$, $r = -.51$). Ratings of stress were also significantly higher for the rotary controller, compared to the touch screen ($z = -2.29$, $p < .05$, $r = -.30$).

There were no significant differences between ratings of interference with the primary task between the touch screen and rotary controller ($z = -1.16$, $p = .14$).

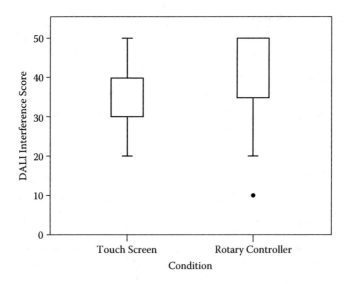

FIGURE 6.14 Box plot of DALI Interference ratings.

For the global attention demand, visual demand, and stress aspects of workload, the participants' ratings supported the secondary task performance results: higher workload would be expected to produce worse performance. Mean ratings of interference for both IVIS were relatively high, indicating that participants thought that both systems affected primary driving performance to some extent; however, they did not perceive a significant difference between the two systems.

Usability Issues

The main aim of this case study was to investigate the usability issues associated with two IVIS input devices, touch screen and rotary controller, so that designers can better understand how to improve the usability of these systems. Usability issues associated with the rotary controller, touch screen, and the design of the GUI in general, together with their causal factors, are presented in Tables 6.6, 6.7, and

TABLE 6.6
Rotary Controller Usability Issues and Causal Factors

Rotary Controller Usability Issues	Causal Factors
Indirect input, increased time for menu scrolling	Input device type (indirect)
Too many options to scroll through before target	GUI/menu structure not optimised
User must check progression towards target	Input device type (indirect)
Visual feedback not strong enough	GUI design not optimised
Unpredictability of movement through alphanumeric sequences	GUI/menu structure not optimised
Lack of rotary controller sensitivity	Hardware issue

TABLE 6.7
Touch Screen Usability Issues and Causal Factors

Touch Screen Usability Issues	Causal Factors
Lack of touch screen sensitivity	Hardware issue
Position of touch screen in relation to driver	Input device type (direct)

TABLE 6.8
GUI/Menu Structure Usability Issues and Casual Factor

Gui/Menu Structure Usability Issues	Causal Factors
Unclear structure of radio submenus	GUI/menu structure design
Unnecessary complexity of air direction task	GUI/menu structure design
Increased time to locate bal/fade menu option	GUI/menu structure design

6.8, respectively. Causal factors distinguish whether the usability issues are attributed to the input device type, GUI/menu structure or a problem of input device/GUI optimisation.

Rotary Controller Usability Issues

Increased task times indicated that it took longer for the rotary controller to scroll between different menu items in order to reach the desired target, compared with moving the hand directly to a target. This was a feature of the device because the translation between controller input and on-screen movement took longer than direct movements of the hand to the touch screen, and also of the menu structure and layout, which, for some tasks, meant that the user had to scroll through a large number of menu options before reaching the target. The latter issue was a problem of GUI/device optimisation because the menus were not designed specifically for navigation with a rotary controller. In a GUI optimised for use with a rotary controller, the number of menu items that a user must scroll through before reaching the target should be minimised. Use of the rotary controller resulted in increased visual demand in order to track the position of the highlight when scrolling between menu items. This was a feature of the input device, because visual target location time would have been the same for both the rotary controller and touch screen. However, with the rotary controller, users had to check the starting and intermediate positions on screen more often. Visual demand was also affected by the visual feedback provided by the GUI, which, in this case, consisted of a yellow box that highlighted each menu item as the user scrolled through them. This was not a particularly strong source of feedback and therefore demanded high visual attention. Results of the SUS also indicated increased frustration among users with the rotary controller. Observations showed that this was mainly restricted to tasks that involved alphanumeric entry. This was because the movement through letters and digits on the GUI was not as logical as it could have been, which produced unpredictability. Again, this could have been improved if the GUI had been optimised for the input type. Frustration was also caused by a slight lack of sensitivity of the rotary controller push-down function.

Touch Screen Usability Issues

The touch screen performed better than the rotary controller in all of the usability measures applied in this study. It was therefore more difficult to identify serious usability issues with this device; however, this does not mean that the touch screen is a perfect IVIS solution. Observations during the empirical tests showed that there was a lack of sensitivity in the touch screen, and this could have been a source of frustration to users. Participants also had to keep their arm outstretched for relatively long periods during the test, although this problem was exaggerated because of the large number of tasks performed in a short space of time in this study.

Graphical User Interface/Menu Structure Usability Issues

There were also a number of task-specific issues, relating to the way certain tasks were presented via the GUI. A higher error rate for the 'play radio' task, supported by observations, showed that it was not clear to users that the Radio 4 preset option was located in a submenu of the AM/FM menu. Observations also showed that the layout of buttons made the 'air direction' task more complex than it needed to be. Users had to work out that buttons needed to be pushed in order to deactivate, as well as activate, different air direction options. This is likely to have increased task time. Users also took a relatively long time to locate the audio balance option, which was one of the factors that increased task times associated with the 'adjust balance' task, compared with other level adjustment tasks, such as 'increase bass'.

Optimisation between the Graphical User Interface (GUI), Task Structure, and Input Device

These results show that there were a number of usability issues associated with the rotary controller, the touch screen, and GUI (which was used for both input devices). Some of these issues can be attributed to the fact that the GUI, and associated menu structure, was not optimised for use with the rotary controller. Whilst these issues can give a good indication of potential problems with IVIS, it is not fair to conclude that these are issues with indirect input devices in general, because in reality a GUI should be optimised for the input device associated with it. The lack of sensitivity found in both the rotary controller and touch screen should be attributed to the specific hardware used in this experiment and is not necessarily a true reflection of the sensitivity of touch screen or rotary controller IVIS used in vehicles today. This is not to say, however, that these issues are not important and designers should always take account of how hardware and GUI/menu optimisation could impact on usability.

The main aim of this paper was to investigate the usability of different input devices, and, therefore, the usability issues of most interest are those that can be attributed to input type. This case study highlighted two usability issues with the rotary controller input device that go some way in explaining why performance was worse with this device, compared to the touch screen:

1. The time required to translate the movement of the rotary controller into movements on screen to reach the target item increases overall task time, compared with moving the hand directly to an option.
2. The visual demand associated with tracking the movement of a highlight or cursor through different menu items is increased, compared with visually locating a target and moving the hand directly there.

These usability issues relate to the nature of input, that is, whether it is direct or indirect. The direct relationship between inputs and outputs is one of the main advantages of the touch screen and evidence of the benefits of this to secondary task interaction times, and primary driving performance has been found in the study presented here. Previous studies have shown that the direct nature of touch screen input increases learnability and initial satisfaction with the device, compared with the rotary controller (Rogers et al., 2005), and this is also supported by the results of the current study.

These findings have shown that it is not only the design of the input device that affects the usability of an IVIS but also the optimisation between the input device and the structure and layout of the GUI. In this study, the GUI was not optimised for use with the rotary controller for some of the tasks. This was because there was a need to have the same content and structure of tasks for both conditions; however, it also means that the results may not be an accurate reflection of performance for some of the rotary controller tasks. These tasks were those that involved some form of level adjustment and those that required relatively long sequences of alphanumeric entry. The rotary controller is ideally suited for level adjustment because the dial can be turned to increase/decrease on a continuous scale; however, in this study the rotary controller could only be used to select a plus/minus button and push down to increase/decrease. It is likely that rotary controller performance would have been improved for these task types had the GUI been designed specifically for this type of operation. The picture is less clear for alphanumeric entry because the rotary controller must still scroll through a large number of menu items in order to select a letter or number and this is a relatively inefficient process. It is likely, however, that a different GUI layout could improve this task for rotary input. For example, in the BMW iDrive, which utilises a rotary dial, letters for address input are arranged in a circle, which represents the movement of the rotary dial more accurately than presenting letters in horizontal lines. Further tests would be required to show if GUI optimisation can improve the performance of the two IVIS.

IMPLICATIONS

The participants had no prior experience of the two devices tested in this case study, and the findings therefore represent the interaction of novice users with IVIS. Nowakowski et al. (2000) reported a 64% increase in IVIS task interaction times with novice users, compared to expert users. This means that it may not be suitable to extend the findings of the current study to users with more experience of IVIS. With experienced users, the rotary controller may demonstrate higher usability according to the measures applied here (Rogers et al., 2005) and this is something that needs to be taken into account in future applications of the evaluation framework. Taveira and Choi (2009) also expressed doubts about the usability of touch screens for older users, particularly in terms of accuracy and comfort. Rogers et al. (2005) reported findings that supported the use of rotary controllers for older adults, as this produced less performance variability, compared with a touch screen. The participant sample used in this study, with a maximum age of 33 and a mean age of 25 years, was not representative of the older driver population. Previous studies (Nowakowski et al., 2000; Tijerina et al., 1998) have found that older drivers take, on average, approximately double the time of younger drivers to perform IVIS navigation tasks, demonstrating an effect of age on IVIS interaction. Tijerina et al. (1998) also found that older drivers produced more centreline crossings and recorded more eyes-off-road time when interacting with IVIS navigation tasks, compared to their younger counterparts. Again, further studies would be needed to evaluate the usability of these two systems with drivers of all ages.

One of the main motivations underlying this work was to help designers identify and understand usability issues. The results show that the empirical methods used in this case study were capable of distinguishing between the two IVIS in terms of primary and secondary task performance, visual behaviour, and subjective usability. It has also been possible to identify a number of serious usability issues based on these results. In order to highlight these usability issues, the driving and secondary task conditions were exaggerated, producing a higher level of demand than would be expected during real driving. It is therefore important that the results, particularly for driving performance and visual behaviour, are interpreted within this context. They provide a relative prediction of usability between the two systems investigated, rather than an absolute measure of the effect of interacting with the two systems on driving performance. This type of empirical testing is therefore recommended for relatively early stages in the evaluation process, when major usability issues still need to be identified with a sample of users. Later in this process, it will be more appropriate to use testing conditions that can replicate real on-road driving more accurately, in order to identify more subtle usability issues and to produce absolute measures of performance.

CONCLUSIONS

Evaluating the usability of IVIS can help designers to understand the limitations of current systems through the identification of important usability issues (Harvey et al., 2011d). In this study, empirical methods were applied in the evaluation of two of the most commonly used IVIS input devices currently used by automotive manufacturers: touch screen for direct input, rotary controller for indirect input. The methods used in this empirical case study make up a detailed, user-centred approach for investigating how different input devices affect performance with particular tasks and GUIs. This has enabled the identification of usability issues that are specific to input device types and also those that are related to other aspects of IVIS design, including GUI/menu structure and hardware characteristics. The usability issues associated with the direct and indirect input device types will be useful to designers who want to select the most suitable device, given the particular task in question. This study has highlighted the difficulty in evaluating input devices independent of GUI layout and menu structure and illustrates the importance of considering the optimisation between input device and GUI/menu structure in design and evaluation. Different input devices are more suited to particular task types, and this points towards a multimodal solution for IVIS. Different menus within a multimodal system will then need to be structured and presented in a way that is optimised for the intended input device in each case.

7 Modelling the Hare and the Tortoise
Predicting IVIS Task Times for Fast, Middle, and Slow Person Performance using Critical Path Analysis

INTRODUCTION

There is a need for usability evaluation at an early stage of product development (Harvey et al. 2011d; Nielsen, 1993; Stanton and Young, 1999a). It is often argued that Ergonomics is involved too late in the engineering design process to have a significant impact, as the 'evaluations' are towards the end of the product development life cycle (Bevan, 1995; Card et al., 1983). Rather, Ergonomics could have much more prominence if it were applied at the beginning of the life cycle to concepts and early prototypes (Stanton and Young, 1999a). This way Ergonomics could guide design in a proactive manner, rather than reacting to poor design at the end of product development, when it is too late to have substantive impact on design (Pettitt et al., 2007; Nowakowski et al., 2000).

MODELLING HUMAN–COMPUTER INTERACTION

Analytic methods are useful for making predictions about the likely usability of products without the need for robust prototypes and user trials, which can often be a costly and time consuming method of testing (Pettitt et al., 2007; Salvucci et al., 2005). An empirical evaluation of different IVIS would take several weeks or months to develop and carry out, whereas an analytic approach allows predictions of IVIS task times to be obtained in a much shorter time (Manes et al., 1997). Analytic methods are suitable for application early in the product life cycle due to their low resource demands (Green, 1999; Kieras and Meyer, 1997); however, there is still an associated cost with HCI modelling, in terms of the knowledge needed to create the models, learning and understanding the theory that underlies a model, and the time taken to generate the models using a particular tool or technique (John and Jastrzembski, 2010) and an aim for the analytic evaluation of IVIS by automotive manufacturers must be to minimise these costs.

TASK TIMES

Task times are a useful measure in evaluating IVIS usability as they provide a quantitative metric of user performance (Baber and Mellor, 2001). Predicted IVIS task times are useful for giving a relative estimate of the time taken to perform different secondary tasks using an IVIS in a stationary vehicle. For example, entering a navigation address usually involves a relatively large number of manual, serial operations, including entering each letter in a town's name into the system. This task is likely to take longer than a task such as 'turn on auto climate', which only involves a small number of total operations to complete. Whilst relative task time predictions give a useful comparison between different tasks and IVIS, it is also important for designers and analysts to be able to predict absolute task times. The 15-second rule has been proposed for the evaluation of in-vehicle secondary tasks (Green, 1999). This rule states that no navigation task involving a visual display and manual controls and available during driving should exceed 15 seconds in duration (Green, 1999; Nowakowski and Green, 2001; Society of Automotive Engineers, 2002). This standard offers a relatively simple guide for assessing the distraction potential of IVIS tasks early in the design process (Reed-Jones et al., 2008); however, it has also been criticised for taking no account of the interruptability of tasks (Reed-Jones et al., 2008), which is likely to reduce its validity as an evaluation technique. Nowakowski and Green (2001) also warned that compliance with the 15-second rule does not guarantee that a task is safe to perform whilst driving and that this practice does not address all the potential causes of driver distraction in the vehicle. It is important that the individual tasks and IVIS are evaluated in detail and on a case-by-case basis to determine their potential to cause distraction. It could be argued that modelling the absolute task times for a particular IVIS/task combination provides a richer level of information and understanding about the tasks than timing task performance and comparing it to a set standard. Breaking tasks down into individual operations can highlight where conflicts are more or less likely to occur with particular driving tasks and this detailed level of insight should further our understanding of what types of interaction have the most potential to cause distraction. The modelling approach should also be simple and relatively quick to perform in order to encourage application by automotive manufacturers as part of the product development process. This approach is investigated in this chapter.

MODELLING TECHNIQUES

There are a number of existing HCI modelling techniques that produce predictions of task time, including Card and colleagues' GOMS (Goals, Operators, Methods, and Selection Rules) technique, and variants of GOMS such as the Keystroke Level Model (KLM); EPIC (Executive Process-Interactive Control; Kieras and Meyer, 1997); ACT-R (Atomic Components of Thought; Anderson and Lebiere, 1998); and, CPA (Critical Path Analysis), which was originally developed as a project network technique (Lockyer, 1984), but has since been applied to model people's activities (e.g., Baber and Mellor, 2001; Stanton and Baber, 2008). GOMS and KLM are based on the Model Human Processor (MHP) proposed by Card et al. (1983), which represents the interactions between our perceptual, motor, and cognitive systems in terms of individual memories and processes. KLM is a simplified version of GOMS (John and Kieras, 1996), and both techniques model behaviour using a sequential ordering of operations (Card et al., 1983). Each operation in the sequence is assigned a time, and total task times are predicted by the model. GOMS and KLM use a small set of predefined operators to build task models, which may limit the application of these methods outside of the desktop computing environment (Byrne, 2001). Because of this sequential ordering of operations in GOMS and KLM, there is no way of representing the overlap between different processing modes (John and Kieras, 1996); however, this issue is addressed by another version of the technique, CPM-GOMS (Cognitive-Perceptual-Motor; John and Gray, 1995; Gray et al., 1993), which is able to model parallel operations in a similar way to CPA. In CPM-GOMS, operators are described at the level of cycle times, as described in relation to the MHP (Card et al., 1983): this is the smallest level of description for operators and therefore requires detailed knowledge of the MHP from the analyst applying the technique (John and Kieras, 1996). As training and application times need to be minimised in any technique proposed for use in the early stages of IVIS evaluation, this knowledge demand could be a significant disadvantage of the CPM-GOMS method. The EPIC architecture is similar to GOMS although it is delivered in the form of a software framework to support computer simulation (Kieras and Meyer, 1997). The technique is capable of modelling detailed mechanisms of information processing and perceptual-motor activity, incorporating a theory of visual attention and perception; however, it requires a powerful programming language to implement, which imposes relatively heavy training and application time demands on analysts (Kieras and Meyer, 1997). In a resource-limited IVIS development process conducted by a typical automotive manufacturer, there is unlikely to be scope for this level of analysis. EPIC has also been criticised for failing to represent a task typical of that which would be performed by the user (Byrne, 2001). A similar cognitive architecture, ACT-R, attempts to address this problem by representing interface objects such as buttons and text objects in order to simulate more realistic user behaviour (Byrne, 2001). Similar to EPIC, application of ACT-R relies on a relatively high level of programming language knowledge, which is a disadvantage given the constraints imposed by the scenario addressed in this study. CPA has its roots in project management (Lockyer, 1984); however, it can be applied to any time-based activity, including the interaction between human and computer (Stanton and Baber, 2008). Similar to other HCI modelling techniques, CPA breaks tasks into operations and assigns times to build a model of total task time. CPA involves the creation of task time diagrams that illustrate the relationships between the individual operations involved in a task. An advantage of CPA over similar methods such as GOMS and KLM is its ability to model parallel operations (Baber and Mellor, 2001). It is also expected that CPA will be simpler to apply than CPM-GOMS, EPIC, and ACT-R, given the constraints of early-stage IVIS evaluation, as analysts do not need a detailed knowledge of programming languages, production rules, or the MHP.

This discussion highlights a cost-benefit trade-off between the various HCI modelling techniques (John and Jastrzembski, 2010). Sophisticated modelling methods such as EPIC and ACT-R produce more rigorous and detailed models but require a highly complicated modelling effort, whereas simpler models such as GOMS and CPA provide less sophisticated functionality but their relative ease of use enables rapid prototyping and requires fewer hours in training and application (Salvucci, 2001). With the constraints of IVIS evaluation, as described in the previous section, in mind, CPA was selected for evaluation of IVIS as it can be applied early in product development, is capable of modelling parallel operations, and has the lowest resource requirements of all the methods reviewed. Although CPA provides a less detailed model of HCI, this simplicity is likely to be an advantage in this context as it will help analysts, that is, the designers and evaluators of IVIS in automotive companies, to understand the structure of tasks and to identify the operations that contribute to high total task times. This information can feed into redesign activities to improve IVIS performance. Kantowitz (2000) suggested that IVIS research must meet the requirements of IVIS designers and the industry in order for it to be useful: this study aimed to develop an evaluation tool with this important need in mind.

CRITICAL PATH ANALYSIS

CPA is used to model the individual operations that are performed in order to complete a task (Baber and Mellor, 2001; Harrison, 1997; Lockyer, 1984). Operations can occur in series or in parallel, and an advantage of CPA is its ability to model both types. Parallel operations can be described according to Wickens' theory of multiple resources, which proposes that attention can be time-shared more effectively between operations in different interaction modes, compared with operations in the same mode (Wickens, 2002). During interaction with an IVIS, the driver must utilize multiple modes in parallel and there is also demand to use the same mode (vision) to monitor both the IVIS and road scene simultaneously. According to the multiple resource model, there will be a conflict between this visual monitoring of two spatially separate scenes, and it will therefore be important for designers to identify the visual operations in an IVIS task, so that potential conflicts can be explored. The CPA technique should support this.

CPA was applied in a previous study (see Harvey and Stanton, IET Intelligent Transport Systems 6(3): 243–258) to predict completion times for nine IVIS tasks. This case study demonstrated the method and was used to assess the usefulness of task time outputs in comparing different IVIS interfaces; however, the predictions were not compared to empirical data and so estimates of validity could not be made. The aim of the study described in the current chapter was therefore to validate the CPA approach and the operation times used to generate the models against empirically derived IVIS task times. In this case, IVIS task times reflect how long it takes the driver to perform IVIS tasks in a single-task environment, that is, just operating the IVIS with no concurrent driving task. This represents the driver's interaction with an IVIS whilst the vehicle is stationary or a passenger's interaction with the IVIS in a moving vehicle. A further aim was to develop a simple software tool to increase the usability of CPA for IVIS designers and evaluators within automotive manufacturing companies.

With the CPA method, designers and evaluators build the model from simple operations, each of which has an associated execution time derived from the HCI literature. CPA diagrams enable designers to study the structure of tasks so that they can learn how this structure influences task times and overall usability. To analyse the tasks that can be performed via one IVIS, CPA is expected to take approximately 2 hours to learn and requires 2–4 hours for data collection and 8–10 hours for analysis. These time estimates were based on an analysis of nine tasks, with a mean of between three and four task steps (see Chapter 5). It is expected, however, that the development of a CPA calculation tool will significantly reduce the time required for analysis and simplify the creation of CPA diagrams. The benefits of a software support tool, in terms of reduced training and application times, have previously been demonstrated for 'CogTool', which was developed to support KLM (John et al., 2004a), and it is expected that similar results will be achieved in this study by automating the CPA process via a spreadsheet tool.

EXTENDING CPA FOR FASTPERSON AND SLOWPERSON PREDICTIONS

In the current study, the CPA technique was extended by developing three versions of the model: fastperson, middleperson, and slowperson. This followed the approach proposed by Card et al. (1983), which would enable the range between best and worst performance to be examined, rather than focussing only on nominal median performance. Card et al. (1983) advised that the range between slowperson and fastperson 'must be kept clearly in mind', so that predictions of the upper and lower boundaries of performance can be used to complement predictions of nominal performance, which will be susceptible to increased inaccuracies due to secondary effects outside the scope of the model. The simplicity of CPA compared to other HCI modelling techniques enabled these fastperson–slowperson operation times to be incorporated into the models relatively easily and the effects of incorporating these ranges to be seen immediately by analysts. The usefulness of fastperson, middleperson, and slowperson values is dependent upon the type of prediction required in a particular circumstance, that is, worst-case scenario, best-case scenario, or an estimate of typical performance (Manes et al., 1997). The procedure used to apply and extend the CPA model is discussed in the following sections.

METHOD

IDENTIFICATION OF OPERATION TIMES

As with the analytic methods case study described in Chapter 5, a set of operation times were derived from the HCI literature: these are shown in Table 7.1. These operations are specific to the touch screen input device that was modelled in this study and so differ from those presented in Chapter 5. There was variation between the times reported for operations by different studies, and these ranges were used to define fastperson, middleperson, and slowperson times for each operation. The final

TABLE 7.1
Relevant Operations Times from HCI Literature and Times Assigned in Current Study

Mode	Task	Time (ms)	Reference	Time Used In Model
Visual	Primed search	1300	Stanton and Baber (2008)	Visually locate single target: 1300 [750~2300]
	Search VDU for new information	2300	Stanton and Baber (2008)	
	Locate target among distractors	750–1400	Wolfe (2007)	
	Recognise familiar words or objects	314–340	Olson and Olson (1990)	Visually locate sequential alphanumeric target: 340 [314~370]
	Perceptual operator (to perceive a single stimulus in the environment)	75–370	John and Newell (1987)	
	Glance at simple information	180	Olson and Olson (1990)	
	Check if on-screen target is highlighted	600–1200	Pickering et al. (2007)	Check if target is highlighted: 900 [600~1200]
Manual	Move hand from steering wheel to dashboard	900 [880~990]	Mourant et al. (1980)	Move hand to touch screen: 900 [880~990]
	Press button/target	200	Baber and Mellor (2001)	Press touch screen target: 200 [112~400]
	Keystroke on a numeric pad	112–400	Card et al. (1983)	
	Key press	120	(John, 1990)	
	Keystroke of a mid-skilled typist	280	Card et al. (1983)	
	Move hand between targets (including press target time)	505–583	Ackerman and Cianciolo (1999)	Homing on target (movement time during visual search assumed extra): 320 [256~443]
		520	Stanton and Baber (2008)	
		368–512	Rogers et al. (2005)	
	Upstroke from key	60	John (1990)	Homing on target: repeat: 60 [50~70]
Cognitive	Make simple selection	990	Stanton and Baber (2008)	Make selection: 990 [660~1350]
	Retrieve item from long-term memory	660–1350	Olson and Olson (1990)	

column of the table lists the times used in the CPA model: middleperson values are shown first, followed by fastperson and slowperson values in square brackets. The assumptions of the CPA model as defined in Chapter 5 are consistent; however, these have been extended for fastperson–slowperson ranges, and so are described again for clarity.

- Time to visually locate a target is 1300 [750~2300] ms, following Stanton and Baber (2008), for any single target and the first alphanumeric target in a sequence.
- Time to visually locate a target is 340 [314~370] ms for any sequential alphanumeric target after the first target in a sequence. It is assumed that users would be more familiar with the layout of an alphanumeric keyboard than with the other menu screens in each system; therefore, search time for alphanumeric targets was reduced.
- No cognitive 'make selection' operation occurs in parallel with a sequential alphanumeric visual search (340 [314~370] ms), following the heuristics for Mental operators devised by Card et al. (1983). Entering a word or telephone number into the system is assumed to be a single 'chunk': users make a decision about the sequence of letters or numbers at the start of the chunk; therefore, individual decisions for each alphanumeric entry are assumed to be unnecessary.
- It is assumed that users move the hand/fingers (touch screen) or the cursor (remote controller) during visual search, even before the target is found (Byrne, 2001). This movement follows the direction of gaze, so only a small 'homing' movement is needed when the target is found (Olson and Olson, 1990). This movement time is not fixed as it varies with the visual search time. It is assumed that the movement starts just after visual search begins; therefore, a value of 1000 [450~2000] ms has been assigned in the models.

A number of further assumptions were made in deciding the range values for each operation in the CPA, as described in the following sections. The fastperson, middleperson, and slowperson operation times for each operation involved in the tasks analysed in this study are also shown in Table 7.1.

VISUALLY LOCATE SINGLE TARGET

Previous studies of visual search have used tasks which range from identifying a number '2' among a set of number '5' distractors (Wolfe, 2007), to a primed search of a Visual Display Unit (VDU) for a train symbol (Stanton and Baber, 2008). In their CPA model of the response time of a rail signaller, Stanton and Baber (2008) used a value of 1300 ms for 'primed visual search' for visual search time for a target that was familiar to the signaller, but its location was unknown. This value was used in the current study to represent the median operation time for 'locate target', as the signaller's action was considered to be very similar to the operation of locating a target on the IVIS when the user knows the name of the target they are searching for but not its specific location on screen. Stanton and Baber (2008) reported a value of 2300 ms to search a VDU for new information, that is, a target which is unfamiliar to the user and its location is unknown. This is used to represent the slowperson user in the current study. Wolfe (2007) reported a range of 750–1400 ms for visually locating a target among a set of distractors: the lower bound is used in the current study to represent the fastperson user.

Visually Locate Sequential Alphanumeric Target

An assumption of the CPA is that the time required to locate an alphanumeric target which is part of a sequence of inputs; that is, the name of a destination or a phone number, will be significantly shorter than the time required to locate a single menu target such as 'radio'. HCI literature reports ranges for recognition of a simple or familiar stimulus of between 75 and 370 ms (John and Newell, 1987; Olson and Olson, 1990). A value of 340 ms was used in the current study to represent the median operation duration, following Olson and Olson (1990). A value of 314 ms was the lower bound for time to recognise a familiar word, reported by Olson and Olson (1990); this was therefore used to represent the fastperson time in the model. A time of 370 ms was the maximum reported in the literature for locating a simple target: this was therefore assigned to the slowperson operation.

Check If Target Is Highlighted

Pickering et al. (2007) reported a range of 600–1200 ms to check if an on-screen target is highlighted. The midpoint of this range (900 ms) was used to represent median (middleperson) operation time, and the minimum (600 ms) and maximum (1200 ms) values in the range were assigned to fastperson and slowperson, respectively.

New Menu

One basic rule for invariance of operation times is that machine-paced operations will always be invariant because they are unaffected by individual differences that occur between human operators. This rule is applied to the 'new menu' operation in the CPA, which describes the transition between two different menu screens, and is controlled by the system rather than the user. A time of 200 ms was estimated for this operation based on the particular system under investigation.

Move Hand from Steering Wheel to Touch Screen

Mourant et al. (1980) reported an average value of 900 ms, with a range of approximately 880–990 ms for time to move the hand from the steering wheel to dashboard-mounted controls. This movement distance is similar to the distance from steering wheel to controller in the current study, and these values were therefore used in the model. A value of 900 ms was used to represent the middleperson time, 880 ms for fastperson, and 990 ms for slowperson.

Move Hand

Move hand time is used to represent the movement of the hand and fingers that occurs in parallel with the visual search operation, that is, whilst a target is being searched for. This movement tends to follow the direction of gaze, so it is assumed that the hand will be close to the target when the target is identified (Olson and Olson, 1990). A further assumption is that this movement will start shortly after the visual search operation, and then both operations will run in parallel; it was therefore assigned a slightly shorter time than the visual search operation: 1000 [450~2000] ms.

Press Touch Screen Target

Card et al. (1983) reported a range of 112–400 ms for a single keystroke. Baber and Mellor (2001) used an average value approximately in the middle of this range (200 ms) to represent a single button press; therefore, this value was used to represent the middleperson value. Times of 112 ms and 400 ms were used to represent the fastperson and slowperson operations, respectively.

Homing on Target

Previous studies have reported times of between 368 ms and 583 ms for physical selection of on-screen targets, combining movement of the hand and pressing a target (Ackerman and Cianciolo, 1999; Rogers et al., 2005; Stanton and Baber, 2008), with Stanton and Baber (2008) assigning an average time of 520 ms to a single button selection. These values include time to press a button, which in this study was estimated at 200 [112~140] ms. Button press time was subtracted from hand movement time to give an estimate of homing time to a target for middleperson (520 − 200 = 320 ms), fastperson (368 − 112 = 256 ms), and slowperson (583 − 140 = 443 ms). Homing time represents the time taken to make the final adjustments to hand and finger position after the position of a visual target has been identified. It is assumed that some movement of the hand and fingers will have already occurred in parallel with the visual search and that this will have positioned the finger in close proximity to the target as physical movement follows the direction of gaze (Olson and Olson, 1990).

Homing on Target: Repeat

Some targets were pressed two or more times in succession, which would have eliminated the homing between targets operation; however, a time for lifting the finger off the button still needed to be assigned. John (1990) used a value of 60 ms for one upstroke in which the user's finger moves upwards, away from the pressed key, in preparation for a second downstroke, that is, the subsequent button press. No other upstroke time data could be found in the literature; therefore, fastperson (50 ms) and slowperson (70 ms) times were estimated either side of the reported value. As this operation has such a small duration and is used relatively infrequently in the touch screen tasks, it is unlikely that slight errors in estimation will have a significant effect on the total task time predictions.

Make Selection

Olson and Olson (1990) reported a range of 660–1350 ms for time to retrieve an item from long-term memory, and Stanton and Baber (2008) used 990 ms to represent time to make a simple selection. These values were assigned to middleperson (990 ms), fastperson (660 ms), and slowperson (1350 ms) in the current study.

DEVELOPMENT OF THE CPA CALCULATOR

The CPA models were originally developed manually, in diagrammatic form, as illustrated in Figure 7.1. This proved to be a very time-consuming process, and a CPA calculator was therefore developed using Microsoft Excel. The calculator was designed to enable fast calculations of task times and to allow the experimenters to instantly see the effects of using different parameters in the model. CPA represents the individual operations that make up a task as a network of boxes, or nodes. The relationships between the operations are defined by arrows that connect the boxes: these show the dependencies between the operations (Harrison, 1997). In a standard CPA diagram, time flows from left to right as the network progresses; however, in the Excel calculator, time flows from top to bottom as this was easier to represent on screen. A screen shot of the calculator is presented in Figure 7.2. Arrows are not displayed in the Excel form; however, the vertical and horizontal positions of the selected cells denote the interrelationships between operations and the operations on the critical path are highlighted in red. Total task time is displayed in the top, right cell. Operation time estimates for fastperson, middleperson, and slowperson were built in to the calculator, and a dropdown menu is used to toggle between the three estimates. Formulae were written to first calculate the EST and EFT for each operation, via the forward pass that moves from top to bottom through the diagram; second, the LST and LFT for each operation, via the backward pass that moves from the 'end' activity at the bottom of the diagram to the starting activity at the top. Next, float time is calculated: all paths through the network with the exception of the

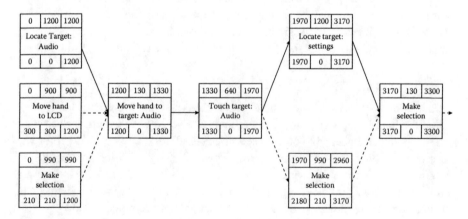

FIGURE 7.1 CPA diagram developed manually, using Microsoft Excel.

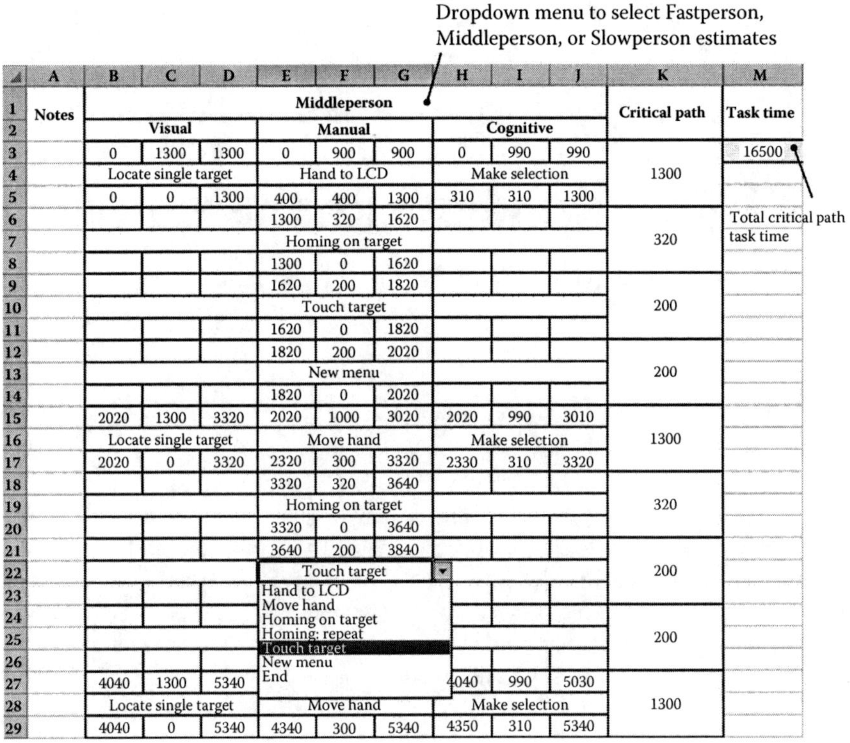

Modelling the Hare and the Tortoise

	A	B	C	D	E	F	G	H	I	J	K	M
1	Notes	\multicolumn{9}{c}{Middleperson}	Critical path	Task time
2		\multicolumn{3}{c}{Visual}	\multicolumn{3}{c}{Manual}	\multicolumn{3}{c}{Cognitive}		
3		0	1300	1300	0	900	900	0	990	990		16500
4		\multicolumn{3}{l}{Locate single target}	\multicolumn{3}{l}{Hand to LCD}	\multicolumn{3}{l}{Make selection}	1300	
5		0	0	1300	400	400	1300	310	310	1300		
6					1300	320	1620					Total critical path
7					\multicolumn{3}{l}{Homing on target}				320	task time
8					1300	0	1620					
9					1620	200	1820					
10					\multicolumn{3}{l}{Touch target}				200	
11					1620	0	1820					
12					1820	200	2020					
13					\multicolumn{3}{l}{New menu}				200	
14					1820	0	2020					
15		2020	1300	3320	2020	1000	3020	2020	990	3010		
16		\multicolumn{3}{l}{Locate single target}	\multicolumn{3}{l}{Move hand}	\multicolumn{3}{l}{Make selection}	1300	
17		2020	0	3320	2320	300	3320	2330	310	3320		
18					3320	320	3640					
19					\multicolumn{3}{l}{Homing on target}				320	
20					3320	0	3640					
21					3640	200	3840					
22					\multicolumn{3}{l}{Touch target ▼}				200	
23					\multicolumn{3}{l}{Hand to LCD}					
24					\multicolumn{3}{l}{Move hand}					
25					\multicolumn{3}{l}{Homing on target}				200	
26					\multicolumn{3}{l}{Homing; repeat / Touch target}					
27		4040	1300	5340	\multicolumn{3}{l}{New menu / End}	4040	990	5030		
28		\multicolumn{3}{l}{Locate single target}	\multicolumn{3}{l}{Move hand}	\multicolumn{3}{l}{Make selection}	1300	
29		4040	0	5340	4340	300	5340	4350	310	5340		

Dropdown menu to select Fastperson, Middleperson, or Slowperson estimates

FIGURE 7.2 Screenshot of the CPA calculator spreadsheet, created in Microsoft Excel.

critical path will have some associated float time. Finally, the critical path is identified and a total task time is calculated. These calculations followed the procedure and formulae outlined in Chapter 5.

COMPARISON OF CPA-PREDICTED TASK TIMES WITH EMPIRICAL DATA

An empirical test was conducted to collect data that would be used to validate the CPA model: the details of the test are described in the following sections.

PARTICIPANTS

Twenty participants (10 male, 10 female) aged between 21 and 33 (M = 25, SD = 2.7) took part in the empirical phase of this study. Participants were recruited via e-mail advertisements, from a sampling frame of University of Southampton Civil Engineering students and staff. Participants were all right-handed. They were each paid £10 for participating. The study was granted ethical approval by the University of Southampton Research Ethics Committee.

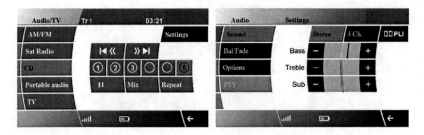

FIGURE 7.3 Screenshots of the play CD and increase bass menu screens (images reproduced courtesy of Jaguar Cars).

EQUIPMENT

Although this study did not involve a driving task, participants were seated in the driving simulator (a right-hand drive vehicle) to complete the IVIS tasks. This was to encourage the participants to use the IVIS in the stationary vehicle in the same way as during driving. A GUI, similar to one used by a major automotive manufacturer, was displayed on a 7 in. LCD touch screen, which was mounted on the dashboard to the left of the driver. The GUI could be controlled by the user, via touching the screen, and also by the experimenter who was seated in the rear of the vehicle, via a laptop. Auditory feedback was not provided by the system. Screen shots of the play CD and increase bass screens are presented in Figure 7.3.

PROCEDURE

At the start of the test, each participant was briefed about the experiment and asked to complete a demographic questionnaire and sign a consent form. Participants were given a 5-minute practice during which they were instructed to use the IVIS to navigate the menus and interact with tasks from each of the function categories (infotainment, comfort, communication, and navigation) in order to familiarise themselves with the interface. The participants were then instructed to perform a set of 14 in-vehicle tasks by the experimenter: these are listed and categorised as audio, climate, communication, or navigation tasks in Table 7.2. Participants were instructed to perform the tasks 'as efficiently as possible', that is, quickly but trying to minimise mistakes.

The order of task presentation was randomised for each participant to minimise practice effects. The 14 tasks were selected to represent the four main IVIS function categories: infotainment, comfort, communication, and navigation. Selection of tasks was based on a number of criteria, including usage whilst driving, frequency of use, and availability in existing systems: see Harvey et al. (2011c) for a detailed discussion of task selection for this study.

DATA COLLECTION AND ANALYSIS

Demographic information was obtained for each of the participants in the study. A key logger recorded secondary task interactions, that is, for each task it recorded which button was activated and at what time. Operation times were calculated for

TABLE 7.2
IVIS Tasks Analysed in the Study

Task Category	Tasks
Audio	Increase bass
	Adjust balance
	Select portable audio
	Play CD track
Climate	Increase fan speed
	Turn-on auto climate
	Reduce seat heat
	Turn-off climate
Communications	Digit dial
	Call from contacts
	Call from calls made list
	Call from calls received list
	Call from calls missed list
Navigation	Enter destination address

fastperson, middleperson, and slowperson versions of the model. Fastperson and Slowperson estimates represent the upper and lower bounds of performance and allow designers to account for variation in performance due to individual differences (Lansdown et al., 2004a). These were then compared to empirical estimates for fastperson and slowperson users, which were represented by the 10th and 90th percentile times from the data. For sample sizes smaller than 20, it is appropriate to use 10th and 90th, as more extreme percentiles such as 5th and 95th, represent the sample range and are likely to be affected by outlying values, therefore producing inaccuracies (Lee, 1986; Snyder et al., 1975; Walter, 1986). The predicted and empirical task times were compared, and the precision (percentage difference between predicted and empirical) was calculated for each task. The results are presented in the following section.

RESULTS

Table 7.3 presents the 10th, 50th, and 90th percentile empirical results alongside the predicted fastperson, middleperson, and slowperson task time predictions. The percentage difference between each prediction and empirical result is also shown.

A difference of 20% is generally considered an acceptable maximum difference between analytic task time predictions and measured task times (Baber and Mellor, 2001; Gray et al., 1993; Pettitt et al., 2007). The CPA calculator predicted middleperson times with an error of no more than 20% for all tasks, with the exception of 'Play CD track' (21.66% difference). There was a mean difference of 8.43% between the middleperson CPA predictions and empirical task times. Three fastperson CPA task times ('Increase bass', 'Increase fan speed', and 'turn off climate') were overpredictions; however, the remaining predictions were all within the 20% difference

TABLE 7.3
Comparison between Empirical and CPA-Predicted Task Times

Task	Empirical Task Times			CPA Predicted Task Times					
	10th%ile	Median	90th%ile	Fastperson	% Difference	Middleperson	% Difference	Slowperson	% Difference
Increase bass	5163	7295	13718	3886[a]	−24.73	5900	−19.12	9156[a]	−33.26
Adjust balance	6740	8548	14759	5490	−18.55	8460	−1.03	13629	−7.66
Select portable audio	4488	6379	9467	4066	−9.40	6260	−1.87	10429	10.16
Play CD track	2603	3485	4672	2748	5.57	4240[b]	21.66	7086[b]	51.67
Increase fan speed	3286	3936	6224	2568[a]	−21.85	3880	−1.42	5813	−6.60
Turn-on auto climate	1606	2234	5935	1430	−10.96	2220	−0.63	3743[a]	−36.93
Reduce seat heat	2868	3776	8507	2568	−10.46	3880	2.75	5813[a]	−31.67
Turn-off climate	1914	2504	3828	1430[a]	−25.29	2220	−11.34	3743	−2.22
Digit dial	10050	13054	21224	9454	−5.93	12460	−4.55	18259	−13.97
Call from contacts	5987	8382	11086	5072	−15.28	7880	−5.99	13172	18.82
Call from calls made list	4547	5403	7345	4066	−10.58	6260	15.86	10429[b]	41.99
Call from calls received list	4085	5954	8978	4066	−0.47	6260	5.14	10429	16.16
Call from calls missed list	4230	5493	9604	4066	−3.88	6260	13.96	10429	8.59
Enter destination address	14646	18898	24844	12090	−17.45	16500	−12.69	24945	0.41

[a] Underprediction, ≤20%.
[b] Overprediction, >20%.

limit, with a mean overall difference of 12.89%. The slowperson model was less precise, with four significant underpredictions ('Increase bass', 'Play CD track', 'Turn on auto climate', and 'Reduce seat heat') and one significant overprediction ('Call from calls made list'). The mean overall difference between the predicted and empirical slowperson times (20.01%) was, however, just within the acceptable 20% limit.

DISCUSSION

CPA has been applied previously in the analysis of a range of tasks including an on-screen menu item selection task (Baber and Mellor, 2001) and rail signaller response (Stanton and Baber, 2008). An issue raised in these studies was how well the CPA could be applied across domains (Stanton and Baber, 2008). In the current study, times from HCI literature were used to model secondary tasks times performed via a touch screen IVIS, producing predictions for fastperson and slowperson users, as well as middleperson, or median, estimates. The majority of predictions were within 20% of the empirical results, which is considered to be an acceptable limit for precision. Based on the total mean precision scores for each of the three models, it appears that the models produced a good estimate of empirical IVIS interaction times. This supports the hypothesis that operation times identified from the HCI literature can be combined to produce precise models of HCI. This finding also supports the use of CPA in this form by automotive manufacturers for quick, early stage modelling of IVIS performance.

The fastperson and slowperson models were less precise than the middleperson model; however, both did produce an overall mean difference between predicted and empirical task times that was within the 20% acceptable limit. At the extreme ends of a distribution, the difference between consecutive percentiles, for example, 98th to 99th, is greater than between the 50th and 51st percentiles in the middle of the distribution (Sivak, 1996). This means that the fastperson and slowperson predictions, which used 10th and 90th percentile values, respectively, were expected to produce less precise predictions than the middleperson (50th percentile) CPA model. Distributions of reaction time data are generally positively skewed because there is a definite lower limit to performance time but there is no upper bound (Chapanis, 1950). The empirical task time data were positively skewed for 11 out of the 14 tasks in this study (exceptions were 'Play CD track', 'Enter destination address', and 'Call from contacts', which were approximately normally distributed). This goes some way to explaining the lack of precision in the slowperson model because the variation in results at the skewed end of the distribution for slow performance would be higher and the model precision would therefore be less likely to fall within the 20% limit. For this reason, a decision was taken to relax the threshold for prediction precision for the slowperson model to 40%. This is illustrated in Figure 7.4 that shows a positively skewed distribution, typical of reaction time data. According to this revised precision threshold, the slowperson model estimated 12 out of the 14 task times with acceptable precision. There is a lack of guidance about acceptable prediction precision in the literature and, although modelling multidimensional scenarios is expected to produce some degree of approximation, it is incredibly difficult to know the magnitude of this approximation (Victor et al., 2009; Pheasant, 1996).

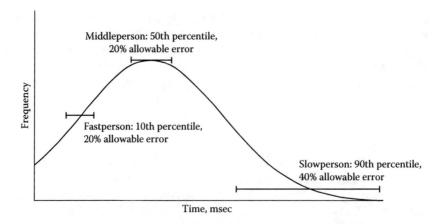

FIGURE 7.4 Positively skewed distribution typical of most tasks in the study, with error bars for fastperson, middleperson, and slowperson.

In this study, the prediction error is also likely to have been increased by the small samples from which the operation times were taken and against which the overall task times were compared (Tilley and Henry Dreyfuss Associates, 2002). Small samples are more likely to be affected by extreme values (Baber and Mellor, 2001), and this would have particularly applied to the slowperson estimates.

Card et al. (1983) reported a large number of times for HCI parameters in their MHP: these were largely secondary data. A brief review of many of the studies from which this data was drawn was undertaken, and this showed that sample sizes for the original data collection varied from approximately 3 to 52 participants (Averbach and Coriell, 1961; Busswell, 1922; Darwin et al., 1972; Fitts, 1954; Fitts and Posner, 1967; Murdock, 1961; Peterson and Peterson, 1959; Sperling, 1960; Sternberg, 1975). Anthropometric data are usually based on much larger samples, for example, height and weight measurements from 10,020 British adults (Rosenbaum and Skinner, 1985), skinfold measurements of 4,049 British businessmen (Richardson and Pincherle, 1969), and height and weight measurements from 13,645 American adults aged 18–74 (Abraham, 1979). An aim for HCI modelling research must be to produce a database of times for interaction parameters that is drawn from a large and representative sample, similar to much of the anthropometric data available today.

APPLICATIONS OF THE CPA MODEL

Static task times have been shown to be a good predictor of eyes-off-road time (Green, 1999), because the more time a driver spends with their visual attention on the secondary task, the less time left for attention to the road scene. The CPA method therefore has the potential to be used to predict the effects of IVIS interaction on driver distraction. Interaction times are an indicator of the efficiency of an IVIS, which is an important factor of usability (Harvey et al., 2011a): the CPA technique should therefore be useful as part of a framework of measures for evaluating the usability of

IVIS (Harvey et al., 2011d). This modelling technique will be useful to designers in the early stages of IVIS/task design to support their understanding of how individual operations are combined to form a task and to highlight operations that have high time demands so that tasks may be redesigned to minimise these operations (Burnett et al., 2004; Llaneras and Singer, 2002; Nowakowski et al., 2000). In this case, slow-person task time predictions represent the limiting user, which in theory could be used as estimates of a realistic maximum task time (Pheasant, 1996); however, the relatively low precision of the slow person model means that any comparisons must be made with caution. It is recommended that the slow person model is used as a first estimate to highlight tasks with potentially high time demands and that these tasks are further investigated using empirical techniques. Again, this reinforces the utility of the modelling technique as one part of a toolkit of methods for evaluating the usability of IVIS: it should not be used in isolation to make design decisions and will never eliminate the need for usability testing with real users (Angell et al., 2006), rather its utility lies in improving designers' understanding of how IVIS design and task structure can affect performance times.

The 15-second rule specifies a maximum recommended time for the interaction with navigation tasks during driving (Green, 1999; Society of Automotive Engineers, 2002). The empirical data showed that two tasks violated the 15-second rule: 'digit dial' according to 90th percentile results, and 'enter destination address' according to 50th and 90th percentile results. These results were supported by the CPA task time predictions: this indicates that the models developed in this study are useful for identifying potentially unsafe tasks at an early stage. This application is an example of the need for not only median, or 'middleperson', estimates but also for predictions of the extreme bounds of performance (Pheasant, 1996). The 15-second rule may, however, not actually be a suitable guide for many IVIS tasks available today because the majority do not need to be completed in a single, uninterrupted sequence. This is dependent on interruptability and resumability of tasks (Burns et al., 2010; Monk and Kidd, 2007; Noy et al., 2004; Reed-Jones et al., 2008), which is influenced by the design of the interface and also the familiarity of the user with the system (Kujala and Saariluoma, 2001). All of the tasks evaluated in this study can be interrupted at various stages and resumed without negatively affecting task completion because the system maintains menu position when a task is interrupted. Although the 'enter destination address' task was predicted to exceed 15 seconds, it could easily be 'chunked' into sections that are less than 15 seconds and separated by glances at the road. A better rule may be to specify maximum interaction times for those 'chunks' of a task that cannot be interrupted without causing a detriment to performance. For example, if a visual search operation has to be interrupted by a glance back to the road, then visual information will be lost in the intervening time and the user will have to spend extra time regaining the lost information when the search task is resumed. In the CPA developed in this study, no effect of resumability was assumed; however, future work is recommended to investigate whether or not this is a suitable assumption to make.

Much of the data collected in this study will be useful in CPA models of different IVIS input types, particularly those which use similar interaction styles to the touch screen, that is, hand movements between physically separated buttons. The literature on interaction styles, such as remote control and speech recognition, will require further investigation to identify specific operation times, for example, the time required to speak certain auditory commands. This may also require empirical studies to investigate new interaction styles, for example, handwriting recognition, which has not been widely studied as an IVIS input method. This work is needed to contribute to a database of HCI operation times, as discussed previously. Application of the CPA model to alternative IVIS input devices will allow quick comparisons to be made between devices at an early stage of product development (Baber and Mellor, 2001; Nowakowski et al., 2000; Pettitt et al., 2007). This will be a useful technique for designers to apply in the earliest stages of product development to help them to understand how different interaction styles will impact on IVIS efficiency and usability, without the need for expensive prototypes.

LIMITATIONS OF THE CPA MODEL

In its current form, the static CPA model does not account for differences in driver characteristics, such as age or experience. The model was based on data from empirical tests with a sample of 20 participants with a mean age of 25 and mean driving experience of 6 years. Participants had no prior experience with the IVIS and were therefore considered novice users. Previous work has found that user characteristics affect secondary task performance, and it is likely that the task time results for older and more experienced drivers would be predicted differently if the model accounted for these factors (Salvucci et al., 2005). Experienced users are likely to be able to perform certain task operations more quickly because they are well practised. For example, users may develop a 'blind touch' for the digit dial buttons in the communication menu (Salvucci, 2001), which would remove the need for some or all of the visual location operations in the CPA model, leaving just cognitive and manual operations in the task (Cockburn et al., 2007). Elderly drivers are expected to exhibit some level of degradation of physiological, sensory, cognitive, and motor abilities (Baldwin, 2002; Collet et al., 2010b; Herriotts, 2005; Lockhart and Shi, 2010), which will affect their interaction with an IVIS, specifically slowing visual search times, selection times, and physical movements between targets. It is likely that all operation times within the CPA model would be increased for older users; however, the extent of this increase is not clear. The fastperson and slowperson versions of the model could be used to approximate the effects of driver characteristics on task time, with the fast model representing experienced users and the slow model representing older users (Jastrzembski and Charness, 2007). Further work would be needed, however, to assess how well these models approximate to the performance of different user groups. It is likely that individual parameters of the models would also need to be altered by different amounts to reflect the complex variations in performance between user groups. For example, task structure may need to be changed in a model of older driver interactions as parallel processing capabilities are likely to be degraded (Hawthorn, 2000).

EXTENSIONS TO THE CPA MODEL

The CPA model provides information about IVIS tasks performed in a single-task situation. Stationary vehicle IVIS task times have been found to have predictive validity for dynamic task duration and glance behaviour (Burns et al.,2010), including eyes-off road time for visual–manual tasks (Angell et al., 2006), and measured static task times produce less variation than task times in a dual-task environment (Burns et al., 2010). However, as static task time does not account for the time-sharing behaviour between primary (driving) and secondary (IVIS) tasks, it can really only be useful for comparisons between tasks, rather than as an absolute measure of distraction potential. The disadvantages of static task time evaluation are exemplified by criticisms of the assumption of the 15-second rule that a task which takes 15 seconds in a stationary vehicle will also take 15 seconds in a moving vehicle (Reed-Jones et al., 2008). Reed-Jones et al. (2008) found that both static and dynamic task times could be used to differentiate between certain tasks; however, static times underestimated the effect of dual-task demands, particularly when hazards in the road were expected by participants. This suggests that task time predictions based on single-task environments, as modelled in the current study, are acceptable for making comparisons between tasks, but not for making absolute predictions about the effects of driving on IVIS interaction. It also assumes that two tasks which are predicted to take the same amount of time to perform will affect the efficiency of performance equally; however, these two tasks may be performed by different IVIS, using different interaction modes, and therefore may produce unequal levels of interference with the driving task. For example, a task performed completely via voice recognition may require the same amount of total time as a task performed using a touch screen; however, the former would involve no eyes-off-road time, whereas the latter would require frequent glances to the touch screen in the interior of the vehicle. According to the multiple resource model, IVIS task involving a larger visual component will be more detrimental to performance because there will be a conflict between visual attention to the IVIS and visual attention to the road (Wickens, 2002). In its current form, the static CPA model only considers the human–IVIS interaction in isolation; however, the focus of Ergonomics research is on the human's performance in a limited domain (Kantowitz, 2000), in this case the driving environment, and there is therefore a need to take a more holistic view of the driver–IVIS interaction. To address the limitations of the static model, the visual behaviour of the driver needs to be integrated into the model to show how different interaction styles affect driving performance. Modelling the division of attention between primary and secondary tasks will provide important information about performance in a dual-task environment. In a dual-task environment, that is, when IVIS tasks are performed at the same time as driving, the utilisation of processing modes will be altered (Wickens, 2002). Driving imposes a high visual load on the driver (Wierwille, 1993), and this will change how visual attention is allocated to secondary tasks. Little is known about how humans balance the attentional demands of primary and secondary driving tasks (John et al., 2004b), and therefore further work is needed to investigate and model IVIS task interactions during driving.

CONCLUSIONS

This study was conducted with the aim of developing the CPA method for predictions of static IVIS task times for fastperson and slowperson users, as well as the average user. Operation times were derived from a review of the HCI literature, and a CPA calculator was developed to facilitate the generation of critical path models. Comparison with empirical task times showed that the fastperson, middleperson, and slowperson models produced total mean differences across all tasks that were within the 20% precision limit, although some individual tasks did exceed this limit, and prediction errors were most numerous for the slowperson model. Accurate time predictions are an important element in the evaluation of IVIS usability, and the CPA model presents a step towards this, which will enable automotive manufacturers to make quick and simple task time predictions at an early stage of product development. The CPA technique allows one element of task performance (task times) to be evaluated without the need for costly prototypes and user trials.

Although the three models produced accurate predictions for the majority of IVIS tasks analysed in this study, the database of operation times upon which the models were based would benefit from extension and validation using larger sample sizes: this would improve the accuracy of the model, particularly for fastperson and slowperson estimates. Dissection of tasks into their smallest components as part of the HTA/CPA process allowed detailed analysis of task structure. These components can be used to generate CPA models for a wide range of alternative IVIS input devices, for example, the visual target location and cognitive selection operations will also be major components of tasks performed by rotary controllers, such as the BMW iDrive, which is an alternative IVIS input style to the touch screen. The successful application of CPA to one of the most widely used IVIS has been demonstrated here, although the current model is limited in that it takes no account of the driving context and the effects of dual-task performance. The model has also been extended for fastperson and slowperson predictions, which provide a useful insight into the range of performance. The CPA technique would be relatively simple to incorporate into early Ergonomics analysis, reducing the need for costly prototypes in the evaluation of task times. Although stationary vehicle IVIS task times can be used to compare the efficiency of different tasks, a dual-task model of the primary–secondary task interactions is needed to provide more information about how attention is managed during driving. This is addressed in Chapter 8 of this book.

8 Visual Attention on the Move
There Is More to Modelling than Meets the Eye

INTRODUCTION

Interacting with an IVIS whilst driving creates a dual-task scenario because the driver must share attention between secondary IVIS tasks and the primary driving tasks. In order for predictions to be made about the usability of IVIS in such contexts, data relating to the way drivers manage the demands between primary (driving) and secondary (IVIS) tasks need to be modelled effectively (Harvey et al., 2011b). The aim of this study was to extend the CPA method for the prediction of task times in a dual-task environment, that is, the time taken to complete secondary tasks whilst undertaking driving tasks simultaneously. This would give an indication of the effect of driving on secondary task interaction times, and this is a factor that affects the usability of an IVIS when used whilst driving.

THE CPA METHOD

CPA was applied in an evaluation of IVIS interactions in a stationary vehicle, that is, tasks performed in a single-task environment, with no concurrent driving tasks (Chapter 7). A CPA model was developed that was capable of predicting the majority of IVIS task times to within 20% of actual measured times. This model of task times was useful for assessing task structure and comparing different IVIS tasks; however, one of the conclusions of the study was that there was a need for a model that could predict IVIS interaction times in a moving vehicle, that is, the time taken to perform IVIS tasks at the same time as driving. In a dual-task driving environment, the utilisation of processing modes will be altered. Driving imposes a high visual load on the driver (Wierwille, 1993), and this will change how attention is allocated to secondary tasks; however, little is known about exactly how humans balance attentional demands in driving (John et al., 2004b). It is proposed that modelling the individual operations involved in interacting with an IVIS whilst driving will lead to greater understanding of the switching of visual attention between primary and secondary tasks.

VISUAL BEHAVIOUR IN DRIVING

The visual mode is the primary information-gathering source used in primary and secondary driving tasks (Wierwille, 1993). Drivers are generally considered to be resource limited with respect to driving (Burnett and Porter, 2001): the driver's attention to the IVIS is inversely related to their capacity for the driving task (Wang et al., 2010) because visual behaviour theories (e.g., Wierwille, 1993) state that the visual resource cannot be divided between spatially separate targets simultaneously. This means that visual attention needs to alternate between the forward road scene (primary driving task) and the IVIS (secondary task) until the secondary task is completed (Sodhi et al., 2002; Wierwille, 1993). With other modes, this division is less clear-cut, and it is more difficult to estimate how nonvisual attention, cognitive in particular, is split between concurrent tasks (Wierwille, 1993). Manual operations tend not to be time-shared in this context because the hands work independently, which means that one hand can remain on the steering wheel (primary task) whilst the other hand is used for the driver-IVIS interaction (secondary task). The same hand can also be used to operate controls mounted on the steering wheel whilst maintaining control of the vehicle. For these reasons, it has been suggested that the visual resource must be given particular emphasis in the design and evaluation of IVIS (Wang et al., 2010; Wierwille, 1993). In this study, it was therefore proposed that combining a model of secondary task performance with a model of visual behaviour would produce accurate predictions of secondary task interaction times in a dual-task driving environment. Development of this dual-task CPA model followed the theory of visual behaviour proposed by Wierwille (1993), in which the driver multiplexes between primary and secondary tasks, starting with a glance at the IVIS task, followed by a return to the forward road scene, followed by another glance at the IVIS task, and so on, until the secondary task is complete. Glance durations at the road and IVIS were measured in a simulated driving environment and these were integrated into the CPA model to represent the visual demands of dual-task performance.

DUAL-TASK GLANCE DURATIONS: PREVIOUS FINDINGS

Pettitt et al. (2007) developed a model of dual-task IVIS performance by integrating the KLM method with glance behaviour patterns used in the occlusion technique. The occlusion technique is used to simulate the allocation of visual attention in a dual-task environment (Senders et al., 1967). Occlusion goggles are used to occlude the driver's vision at regular intervals, to replicate the glance behaviour of a driver looking at an IVIS or at the road during the driving task (Burnett et al., 2004). The technique uses durations of 1500 ms for each vision and occlusion interval (International Organization for Standardization, 2007). These glance durations are based on data reported in the literature (see, e.g., Baumann et al., 2004; Green and Tsimhoni, 2001; van der Horst, 2004) and represent the maximum tolerable time for glances at the IVIS and road, rather than typical (i.e., average) glance times (Baumann et al., 2004; Lansdown et al., 2004b).

The occlusion technique simulates the visual sampling technique proposed by Wierwille (1993). This model proposed that drivers' in-vehicle glances are between

600 and 1600 ms in duration. Wierwille suggested that drivers tend to return to the forward scene after 1000 ms or less if the information presented on an in-vehicle display can be chunked in this time. If not, he suggested that drivers will continue the glance at the IVIS for up to 1600 ms as a maximum limit before time pressure and uncertainty about the forward scene force a glance back to the road. Wierwille (1993) warned that models based on mean glance duration should be interpreted with caution as there is often a great deal of variation in the data. He found that some drivers' glances inside the vehicle were up to 2 s longer than those of other drivers. Previous studies of visual behaviour have also demonstrated wide variations in glance durations, which have been attributed to differences in user characteristics (Wierwille, 1993), environment (Senders et al., 1967), task structure and device type (Mourant et al., 1980). For example, Mourant et al. (1980) found that glance frequencies and duration changed according to the distance the hand moves to a control and the type of visual information being gathered. These early studies that provided the glance data upon which Wierwille's sampling model was based used dashboard and stalk controls, in contrast to today's IVIS, which consist largely of screen-based interfaces located in closer proximity to the driver's field of view. This is likely to have an effect on times, with a reduction in glance durations expected for visual interfaces that are closer to the forward road view (Wierwille, 1993); it is therefore expected that glance times reported in more recent studies would be reduced. These factors make it difficult to predict glance durations for an IVIS that incorporates a large number of different screens and menu items and is used by a wide variety of drivers, under varying conditions. A goal of the case study described in this chapter was therefore to identify the glance durations that were appropriate for modern IVISs, accounting for the effects of the associated tasks, users, and environment, and to investigate whether these values could be incorporated into the CPA model for accurate dual-task IVIS time predictions.

METHOD

The visual behaviour data were collected in the empirical driving simulator case study described in Chapter 6. Originally, the data were used to compare the percentage of time participants spent looking at the road compared to the IVIS. In this study, the data were reanalysed in order to calculate median glance times to the road and LCD. The participant sample, equipment used, and experimental procedure were described in Chapter 6. In this study, only 14 tasks out of the original set of 17 were analysed: this was because of the variation in task structure in three of the tasks, as discussed in Chapter 7. The procedure for the reanalysis of the glance behaviour data is described in the following section.

GLANCE BEHAVIOUR DATA ANALYSIS

In the empirical study, task time data was collected by a keylogger, which recorded each button press and the corresponding time. Visual behaviour was monitored throughout the trial using an eye-tracking system that logged the participants' gaze locations at 30-ms intervals. The locations were categorized as LCD touch screen, left projector screen, middle projector screen, right projector screen, and instrument cluster. Any glance at one of the three projector screens was categorized as a glance at the road scene. The visual behaviour data for four of the participants was removed due to poor tracking accuracy, as described in Chapter 6.

The visual attention data were analysed to identify individual glances at the road scene and LCD screen. Glance behaviour is useful in the prediction of in-vehicle task times (Sodhi et al., 2002). A glance is defined as the period of visual attention in which a driver receives visual information from either the forward road scene or the IVIS LCD screen (Sodhi et al., 2002). In this study, the duration of a glance was measured from the time at which the direction of the gaze moved toward a target until the time at which it moved away (International Organization for Standardization, 2007). The glance data were also filtered to remove saccades and extended glances. A saccade is a brief movement of the eyes between visual fixations, during which no visual information is encoded (International Organization for Standardization, 2002; Salvucci, 1999). Saccades were defined as any glances with duration of 100 ms or less (Horrey and Wickens, 2007; Salvucci, 1999). Extended glances were defined as glances that were longer than 2 s in duration (International Organization for Standardization, 2002). Studies have shown that drivers do not generally glance away from the road for more than 2 s (Victor et al., 2009; Alliance of Automobile Manufacturers, 2006; Horrey et al., 2006). Extended glances may have been recorded at the start or end of the driving trials, when the participant's attention did not need to be on the road scene. These were removed as this study was focussed on glances that occurred when the participants were performing secondary tasks. Glances that occurred outside of secondary task performance, for example, before the instructions to start a task had been given, were also excluded from the analysis.

DEVELOPMENT OF A CPA MODEL FOR DUAL-TASK IVIS INTERACTION

GLANCE BEHAVIOUR DATA

Glances were expected to switch between the projector screens on which the driving scene was displayed and the LCD touch screen during IVIS interaction. Median glance times were calculated from the empirical driving study data: these were 430 ms to the LCD screen and 687 ms to the road scene.

MODEL ASSUMPTIONS

To investigate whether the visual behaviour model derived from the empirical results produced precise predictions of dual-task IVIS interaction times, the glance patterns were integrated into the CPA model that was developed for stationary IVIS interactions, as described in Chapter 7. In the development of the dual-task CPA model, a number of important assumptions were made.

Visual Attention on the Move

THE VISUAL MODE IS MOST IMPORTANT DURING DRIVING

There is a relationship between eye movements and attention (de Winter et al., 2009; Shinar, 2008). The visual mode is the main mode of information presentation during driving, and it should therefore be possible to infer a great deal of information about primary and secondary task performance by analysing visual behaviour (Wierwille, 1993).

VISUAL INFORMATION IS PROCESSED IN 'CHUNKS'

All visual operations will be subject to interruptions if the period of constant visual attention required for an IVIS operation exceeds the maximum glance time at the LCD. In the empirical study, the median glance time at the LCD was 430 ms. This indicates that visual information from the IVIS can only be received in 'chunks' of 430 ms or less. For visual IVIS operations that exceed this limit, the driver will have to look back at the road and then return his or her gaze to the IVIS. The switching of visual attention between the IVIS and road scene and the effect of this on operation time according to the median glance times is illustrated in Figure 8.1.

The top and bottom portions of the diagram show the same task segment, which consists of a visual search operation, a manual 'move hand' to LCD operation, and a cognitive 'selection' operation. The top portion is the stationary task segment, performed in a single-task environment. The bottom portion shows the dual-task segment, in which the visual operation is split when the driver glances away from the IVIS: this increases the overall operation completion time. In the dual-task scenario the simultaneous manual and cognitive operations have increased 'float' time because they can occur anytime during the visual operation time. When the visual

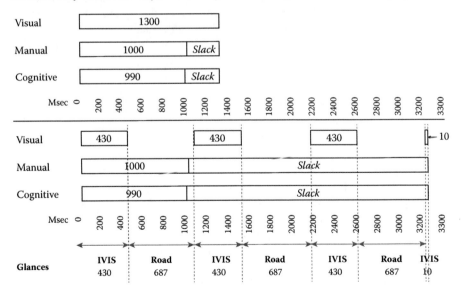

FIGURE 8.1 Change in visual operation time during dual-task IVIS interaction with switching of visual attention between the road and IVIS.

glance times are added to the entire task, there will be an increase in the overall task time. All operations that do not require vision can be performed whilst the driver's visual attention is on the road ahead and is not subject to interruptions from the primary task. Figure 8.1 shows the nonvisual operations continuing into the eyes-on-road period, because they do not require visual attention to complete. At times during the task when there are no visual operations, the glance behaviour will have no effect on task completion time because the operations can be completed nonvisually.

Visual IVIS Operations Cannot Start during On-Road Glances

IVIS operations requiring vision cannot start during a period when the driver's visual attention is directed toward the road ahead. The visual behaviour data showed that median glances to the road lasted 640 ms. It is assumed that when the driver's attention is on the road, it generally stays there for this period of time. This means that new visual operations cannot start during an on-road glance because there is a need to give the road visual attention for a certain length of time.

Median Glance Times Are Most Representative for All Versions of the Model

Median glance times were assumed for the fastperson, middleperson, and slowperson versions of the CPA model. The 10th and 90th percentile glance times could have been used for the fastperson and slowperson models, respectively; however, it was not appropriate to assume that these values would have been associated with the best and worst levels of performance. This is similar to the problem of adding two anthropometric 90th percentile values, for example, to infer the value of a greater part: they will not necessarily add up to the 90th percentile value of that greater part (Bullinger and Dangelmaier, 2003). It may be the case, for example, that participants who were more willing to take risks demonstrated 10th percentile glances to the road scene with 90th percentile glances to the LCD; however, the same participants would not necessarily exhibit 10th or 90th percentile task interaction times.

Dual-Task CPA Calculator

A CPA calculator was developed using Microsoft Excel as part of the single-task CPA modelling work (see Chapter 7). This allowed the experimenters to select different operations from drop-down lists; the form would then display an associated time and would calculate all the components of the CPA (EST, EFT, LST, LFT, Float Time, and Critical Path Time). The visual behaviour rules were built in to the CPA calculator to enable predictions of dual-task IVIS interaction times. The dual-task CPA calculator applied the glance switching behaviour rules to any operations that required vision to calculate the effect of integrating the glance times into the model. Rules were also applied to nonvisual operations that could occur at the same time as glances at the road. Task times were calculated using the dual-task CPA calculator.

RESULTS

The dual-task CPA calculator was used to predict interaction times for the set of touch screen IVIS tasks. These were compared to the task times measured in the empirical study. A limit of 20% prediction error was defined for the fastperson and middleperson predictions based on accepted thresholds used in previous studies (Baber and Mellor, 2001; Pettitt et al., 2007). This limit was extended to 40% for the slowperson predictions because the nature of the positively skewed task time distribution resulted in task times at the right end of the distribution being more variable than those toward the middle and left of the distribution. The setting of these precision limits was discussed in Chapter 7. The CPA predictions are presented in Table 8.1.

For the middleperson task times, the dual-task model estimated just one task time with an error of 20% or less. All predicted times were overestimates of the empirical task times. The average prediction precision for the middleperson estimates was 56.10%, which is well outside the acceptable limit. All 14 task times predictions for the fastperson model were outside the acceptable 20% limit, and like the middleperson model, all were overpredictions. The fastperson model produced an average error of 87.55%. The slowperson model estimated 7 out of the 14 task times within 40% precision. Although the slowperson model produced the highest number of precise predictions, the average precision (44.03%) was still outside the required limit for this model.

The model made a number of large overpredictions of task times for the middleperson, and the precision was poor for the majority of the tasks analysed according to all three models. To investigate the causes of these prediction errors, glance behaviour data from the empirical test was studied in more detail. Facelab software was used to produce a video of the participants' interactions with the IVIS with a gaze circle overlaid on the image to indicate the target of visual attention in real time. Visual behaviour profiles were also created to plot the gaze coordinates against time for each task. Owing to the large volume of data that needed to be analysed per task, this in-depth analysis was performed for a single participant from the study. The participant was selected because her glance times were similar to the median glance times, which were calculated for the whole participant sample, and was therefore a representative case for further analysis.

CASE STUDY: GLANCE BEHAVIOUR IN A DUAL-TASK ENVIRONMENT

The aim of this case study was to extract potential regularities in glance patterns that would develop the theory of visual behaviour to a level more representative of real-world, dual-task IVIS interaction. The analysis of data from a single participant enabled this improved focus (Hancock et al., 2009a) to understand visual processing at a very detailed level.

SHARED VISUAL ATTENTION

The video data showed that the participant moved their head slightly towards the LCD during the tasks; however, the gaze circle moved frequently back towards the road for short glances. This supports the glance switching behaviour that was built into the dual-task model. The video data also showed that the participant seemed to be

TABLE 8.1
Empirical Task Times Compared with Predictions from the CPA Model

Task	Empirical Task Time				MCPA Times					
	10th%ile	Median	90th%ile	n	Fastperson	% Difference	Middleperson	% Difference	Slowperson	% Difference
Increase bass	4817	7932	15903	17	9395[a]	95.04	13971[a]	76.13	21626	35.99
Adjust balance	9774	13199	41351	17	13836[a]	41.56	19713[a]	49.35	29865	−27.78
Select portable audio	5305	10565	16014	20	9888[a]	86.39	13791[a]	30.53	23343[a]	45.77
Play CD track	3903	5113	7544	20	6821[a]	74.76	9423[a]	84.29	15791[a]	109.32
Increase fan speed	2582	4311	8409	20	6328[a]	145.08	9603[a]	122.76	14074[a]	67.37
Turn on auto climate	1883	4407	8347	20	3754[a]	99.36	5055	14.70	8239	−1.29
Reduce seat heat	3369	5909	16061	17	6328[a]	87.83	9603[a]	62.51	14074	−12.37
Turn off climate	2117	3996	7697	17	3754[a]	77.33	5055[a]	26.50	8239	7.04
Digit dial	12778	21371	30313	9	23832[a]	86.51	30696[a]	43.63	44826[a]	47.88
Call from contacts	5819	11216	36103	13	12268[a]	110.83	18159[a]	61.90	30208	−16.33
Call from calls made list	5498	8182	12228	19	9888[a]	79.85	13791[a]	68.55	23343[a]	90.90
Call from calls received list	6454	10015	15475	20	9888[a]	53.21	13791[a]	37.70	23343[a]	50.84
Call from calls missed list	4677	8181	13329	19	9888[a]	111.42	13791[a]	68.57	23343[a]	75.13
Enter destination address	16975	28516	46682	14	29966[a]	76.53	39432[a]	38.28	59930	28.38

[a] Over-prediction (>20% for middleperson and fastperson; >40% for slowperson).
[b] Under-prediction (≤20% for middleperson and fastperson; ≤40% for slowperson).

Visual Attention on the Move

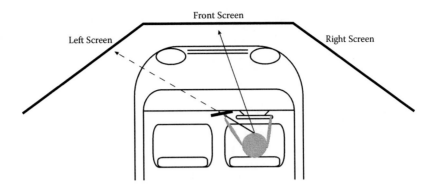

FIGURE 8.2 Simulator schematic showing the position of the LCD in relation to the left projector screen.

able to successfully perform the primary driving tasks, in many cases, whilst their visual attention was directed toward the IVIS. This indicated that the participant was managing her visual attention to the road and LCD by monitoring the road scene and IVIS display simultaneously. The participant also made a relatively high number of glances to the left projector screen, but not to the right. In the earlier visual behaviour model, left screen glances were categorised as glances at the road scene to obtain primary driving information only; however, the video data showed that during glances at the left screen the participant was able to continue to perform the IVIS task. A possible explanation for this behaviour could be that because the LCD screen was located directly in front of the left projector screen from the driver's point of view, the participant was using glances at the left screen to obtain some road information whilst maintaining some visual attention to the LCD. Figure 8.2 illustrates the approximate configuration of the driving simulator, projector screens, and position of the LCD screen inside the vehicle. This shows the proximity of the LCD screen and left projector screen, which may have encouraged drivers to use 'shared glances' to monitor both targets simultaneously. Shared glances involved the participant fixating on a point on the left projector screen, but in proximity to the IVIS. It is very difficult to locate the target of attention from eye position alone (Fleetwood and Byrne, 2006; Salvucci, 2000); however, a split in attention between the road and IVIS is inferred in this case because the participant was clearly able to perform the driving and secondary IVIS tasks simultaneously when engaging in shared glances. To investigate this behaviour further, the participants' glance profiles for each task were also analysed: the profile for the 'adjust balance' task is shown in Figure 8.3.

The graph shows the y-axis gaze position against time. The y-axis position was chosen as it clearly differentiated glances at the LCD and left screen as the participant tended to focus just above the LCD screen during what are hypothesised to be 'shared glances'. These shared glances are shown by the shaded areas in the gaze profile. The 'adjust balance' task consisted of six menu target selections, which are shown by the dashed lines on the graphs: these lines mark the point in the task at which each target was selected by a physical press on the screen. The plot illustrates the glance switching behaviour between the road scene and LCD. The participant also

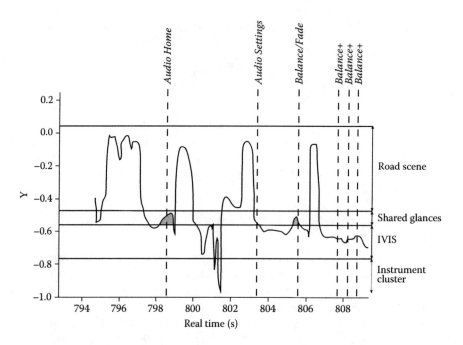

FIGURE 8.3 The case study participant's gaze profile for the 'adjust balance' task.

glanced once at the instrument cluster to check the speedometer (around 801.5 s). All target selections occurred when the participant's gaze was directed towards either the LCD screen or left projector screen. Between approximately 797 s and 799 s, and 803 s and 806 s, the participant's gaze alternated between the LCD and left screen. This could indicate that they were making brief glances to the left screen to obtain road information whilst simultaneously monitoring the IVIS display. In both of these sections, the participant made successful interactions with the IVIS, and this supports the hypothesis that they were able to monitor the IVIS whilst looking at the road. It could be that the participant used this shared glance behaviour, that is, monitoring the IVIS and road screen simultaneously, in preference to making more frequent glances back to the middle and right projector screens for road information. The shared glance behaviour may have allowed the participant to obtain enough information about both the primary and secondary tasks to successfully press each separate target; however, the graph shows that the participant tended to look back to the road scene via the front and right projector screens after a target had been selected in order to obtain more detailed information about the road environment.

This finding prompted the development of a revised glance behaviour model, which classified all glances to the left projector screen as 'shared glances', in which information from both the road and IVIS was being obtained and primary and secondary tasks were being performed in parallel. This shared glance model suggested that visual attention can be shared to some extent between primary and secondary tasks when the visual information necessary for both is in proximity. A number of assumptions were made to support the model.

Shared Glances Are Used to Obtain Visual Information from the LCD and Road Simultaneously

Glances at the left screen are used to obtain information from both the LCD screen and the road scene because at this point information from both sources is in close proximity. These are shared glances in which visual attention is shared, relying on some peripheral monitoring of either the IVIS or road. All glances to the LCD screen are used to obtain information from the IVIS only, and all glances to the middle and right projector screens are used to obtain information about the road scene only.

Visual Information Is Chunked into a Sequence Consisting of One IVIS Glance–One Shared Glance–One IVIS Glance, before Visual Attention Reverts to the Road Scene

In the model, a shared glance and a further IVIS glance are added after a glance to the IVIS if the visual component of that particular task step, that is, locate target and move hand to target, is incomplete after the first IVIS-only glance. An example of this sequence is shown in Figure 8.3, between approximately 798 and 799 s, where the participant glances at the LCD, followed by a shared glance (shaded area) and a second glance at the LCD, before looking at the road scene. If the visual component is still not complete after this IVIS glance—shared glance—IVIS glance sequence, then visual attention is diverted to the road, and the secondary task will only be resumed after the road glance. The median glance times for the shared glance model are shown in Table 8.2.

The effect of this shared glance model on visual attention during task interaction is illustrated in Figure 8.4. In the shared glance model, the visual operation (1250 ms in total) is completed in just two 'chunks', consisting of a combination of IVIS and shared glances. This reduces the number of glances at the road scene and total task time, in comparison with the earlier visual behaviour model.

GLANCE BEHAVIOUR FOR SEQUENTIAL OPERATIONS

The gaze profiles for the case study participant also showed that sequential operations, where all targets are displayed on the same menu screen, for example, entering

TABLE 8.2
Median Glance Times for the Shared Glance Model

LCD Screen	Road Scene	Shared View
\multicolumn{3}{c}{Glance Time (ms)}		
430	750	360

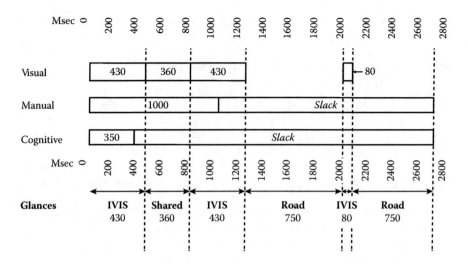

FIGURE 8.4 Effect of shared glances on the model of IVIS interaction.

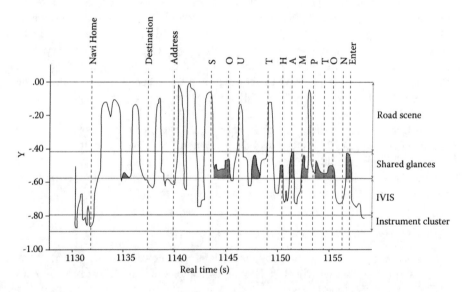

FIGURE 8.5 The case study participant's gaze profile for the 'address entry' task.

letters in an address, resulted in a different pattern of glance behaviour than discrete menu selections, for example, selecting a single option on a navigation submenu. This is illustrated in the gaze profile in Figure 8.5.

Each of the discrete menu selections (Navi Home, Destination and Address) is followed by at least one glance back to the road scene; however, the sequential letter entry options are performed in 'chunks' of between one and four letters in length (average two letters) during which the participant's gaze is directed toward the LCD or shared glance region. Each of these chunks is followed by a glance at the road.

This glance behaviour was built into the shared glance model: a glance at the road was added to the critical path after each chunk of two sequential operations. The shared glance model was also integrated into the CPA calculator by replacing the IVIS-only glance from model 1 with the IVIS glance—shared glance—IVIS glance sequence. Additional formulae were created to identify sequential or repetitive operations in a task and apply the two-letter chunking rule described above.

RESULTS: SHARED GLANCE CPA MODEL

The fastperson, middleperson, and slowperson predictions from the shared glance model were calculated using the revised CPA calculator (see Table 8.3) and compared to the empirical task times and to the results of the earlier model.

Ten out of the fourteen middleperson task times were predicted to within 20% precision with the shared glance model, and the average difference was 16.42%. The remaining four task times ('Play CD track', 'Increase fan speed', 'Call from calls made', and 'Call from calls missed') were overpredicted by the model. The 'Play CD' task was also overpredicted by the single-task model (see Chapter 7), and this error was carried over into the dual-task model. The error in the predictions of the 'Call from ...' tasks could be attributed to anomalies in the empirical data, as the 'Call from calls received' task, which is identical to the other 'Call from ...' tasks, was predicted within acceptable precision limits. The fastperson and slowperson models also produced overpredictions for the 'Call from calls made' and 'Call from calls missed' tasks, but not for the 'Call from calls received' task, lending support to this explanation; however, it is very difficult to identify the exact cause of the error without conducting further comparisons on a wider range of tasks and with a larger sample of participants for the empirical tests. The fastperson model produced predictions within acceptable limits for eight out of the 14 tasks, which is a marked improvement on the previous dual-task model (see Table 8.1). The average error for the fastperson predictions was 22.29%, which was just outside of the acceptable limit. The fastperson model overpredicted task times for task involving repeat target selections, but not sequential target selections, suggesting that the rules for these types of activities may need to be revised. For example, the results show that tasks involving repeat target selections are completed quicker than predicted by the model; therefore, it may be likely that participants are able to perform more than two repetitive operations in series before glancing at the road. The slowperson model also produced eight task times predictions within the acceptable limit for precision (40%), and the task time estimates had an average difference of 25.33, which was well within the precision limit. The majority of slowperson task times estimates were underpredictions, suggesting that the dual-task environment is generally more detrimental to task performance than currently reflected by the shared-glance model. The three largest underpredictions for the slowperson model were for the tasks 'Adjust balance', 'Reduce seat heat', and 'Call from contacts'. Examination of the empirical data from these tasks shows that the 90th percentile tasks times were disproportionately larger than the 50th percentile times, compared with the rest of the tasks studied. This could reflect anomalies in the empirical data or might indicate that the tail of the distribution of task times for these tasks is longer than expected due to

TABLE 8.3
Empirical Task Times Compared with Predictions from the Shared Glance Model

	Empirical Task Time					MCPA Times				
Task	10th%ile	Median	90th%ile	n	Fastperson	% Difference	Middleperson	% Difference	Slowperson	% Difference
Increase bass	4817	7932	15903	17	6212[a]	28.96	9540	20.27	13886	-12.68
Adjust balance	9774	13199	41351	17	9468	-3.13	14160	7.28	20379[b]	-50.72
Select portable audio	5305	10565	16014	20	6018	13.44	10110	-4.31	15729	-1.78
Play CD track	3903	5113	7544	20	4262	9.20	6990[a]	36.71	10736[b]	42.31
Increase fan speed	2582	4311	8409	20	4456[a]	72.58	6420[a]	48.92	8893	5.76
Turn on auto climate	1883	4407	8347	20	2506[a]	33.09	3870	-12.19	5743	-31.20
Reduce seat heat	3369	5909	16061	17	4456[a]	32.26	6420	8.65	8893[b]	-44.63
Turn off climate	2117	3996	7697	17	2506	18.38	3870	-3.15	5743	-25.39
Digit dial	12778	21371	30313	9	14718	15.18	18210	-14.79	24159	-20.30
Call from contacts	5819	11216	36103	13	7774[a]	33.60	12480	11.27	19972[b]	-44.68
Call from calls made list	5498	8182	12228	19	6018	9.46	10110[a]	23.56	15729	28.63
Call from calls received list	6454	10015	15475	20	6018	-6.76	10110	0.95	15729	1.64
Call from calls missed list	4677	8181	13329	19	6018[a]	28.67	10110[a]	23.58	15729	18.01
Enter destination address	16975	28516	46682	14	18230	7.39	24450	-14.26	34145	-26.86

[a] Overprediction (>20% for middleperson and fastperson; >40% for slowperson).
[b] Underprediction (≤20% for middleperson and fastperson; ≤40% for slowperson).

particular difficulties faced by some participants when performing these tasks. There is however, no common characteristic present in these three tasks that explains why some participants would have found them particularly difficult to perform.

DISCUSSION

Two models of visual behaviour have been proposed and tested in this study. The second model, which classified glances as IVIS-only, road scene-only, and shared between IVIS and road, resulted in the most precise predictions of task time for fastperson, middleperson, and slowperson estimates. This result lends support to the shared glance hypothesis as a more suitable model of visual behaviour in a driving/IVIS interaction scenario, compared with a model that treats glances at the road and IVIS as distinct episodes of visual attention. There were, however, still some errors in the predictions from the shared glance model, demonstrating that this is still very much a work in progress and that further investigation is required into visual behaviour whilst driving. One of the major causes of prediction error is likely to have been the small sample sizes on which the models and comparisons were based (Baber and Mellor, 2001). This will have affected the data at the tails of the task time distributions more than the results that were based on median values.

Multidimensionality

The CPA model developed in this study is a multidimensional model, incorporating visual, manual, and cognitive secondary task interaction activities and a pattern of visual behaviour based on dual task performance in a driving environment. In theory, there are a huge number of factors that affect task performance; however, there are likely to be just a few parameters that are responsible for most of the variability in IVIS task times (Victor et al., 2009). In this study, glance patterns were used as one of the major model parameters because the visual mode is most widely used during driving (Wierwille, 1993), and visual attention times are strongly linked to driving performance (Wang et al., 2010). Having said this, other factors such as driver competencies (Stanton et al., 2007a), environmental effects (International Organization for Standardization, 1996; Senders et al., 1967), and training (Llaneras and Singer, 2002; Commission of the European Communities, 2008) will have influenced the precision of the predictions against the empirical results. Based on the experience gained from the development of the CPA models, however, integration of further dimensions of human performance should be attempted with caution. A comprehensive model of human–computer interaction is seen by many as a 'holy grail' (John, 2011); however, the collective effect of multiple interacting dimensions is extremely difficult to control and the sensitivity of humans to different system factors is likely to magnify this effect (Meister, 1989; Schoelles and Gray, 2000). It might also be the case that task structures change in a driving situation compared with interaction with an IVIS in a stationary vehicle. For example, the driver may need to perform more visual checks on the LCD after interactions are interrupted by the primary task, in order to reacquire information (Nowakowski et al., 2000).

Further investigation is necessary to develop rules to guide the structuring of tasks according to the situation in which the interaction is taking place.

There were three cases, all fastperson tasks, where the empirical dual-task IVIS interaction time was shorter than the empirical single-task IVIS interaction time: 'Increase bass', 'Increase fan speed', and 'Call from contacts'. Tsimhoni et al. (1999) also identified cases in which dual-task times for short tasks were actually shorter than single-task times for the same tasks. This effect has been attributed to an increase in pressure to complete the tasks whilst driving in order to return attention to the primary task (Tsimhoni et al., 1999; Tsimhoni and Green, 2003). This effect would be incredibly difficult to account for in the CPA model as it would be highly dependent on the situation and the characteristics of individual drivers. To minimise these effects, a larger sample size is required, and this should be an aim of future work, in order to develop the models and the database of operations upon which they are based.

IMPLICATIONS FOR VISUAL BEHAVIOUR THEORY IN DRIVING

The analysis in this study has suggested that rather than vision being a single resource that cannot be time-shared, visual attention can be divided to some extent between the road and the IVIS. It is likely that some drivers are able to use peripheral vision to monitor the road scene, whilst maintaining some visual attention on the LCD, enabling interaction with secondary tasks during these shared glances. This will be dependent on the location of the IVIS, relative to the driver's useful field of view (UFOV) (Rog et al., 2002). The division of focal and peripheral attention means that visual IVIS operations would not completely stop during some glances to the road; rather, they may slow down or alter in some way. This influenced the way in which glance behaviour was allocated in the dual-task CPA model; however, as the precision of the predictions varied between tasks, it appears that certain operations may be more suited to peripheral visual processing than others. Based on the results of this study, it is very difficult to identify the type of information that can be processed during shared glances. It is possible, however, to make some predictions about the nature of the information that can be processed during shared glances, based on our knowledge of peripheral visual processing. The visual periphery has low acuity, compared with focal vision; however, it is capable of detecting movements and changes in visual displays (Wittmann et al., 2006; Kramer and McCarley, 2003). Peripheral vision is involved in 'pre-attentive processing', which extracts simple features from a scene (Fleetwood and Byrne, 2006), to determine whether a 'serial search' is needed to detect particular elements using focal attention (Snowden et al., 2006; Wolfe, 1998). Based on this, it is suggested that there are certain aspects of the IVIS tasks that are monitored via this pre-attentive processing, using peripheral vision. These are likely to include monitoring the position of the current highlighted target and position of the finger over the screen, and detecting the appearance of a new menu screen, a highlighted target, or alphanumeric entry. Whilst pre-attentive processing is not responsible for identification of visual targets, it is likely to be useful in guiding the driver's attention to particular targets (Wolfe, 1998), which may reduce the total visual search times involved in IVIS interactions.

The plots of glance behaviour showed that sequential operations could be performed with less frequent glances at the road scene than discrete target selections, and this supports the hypothesis that peripheral monitoring is useful to some extent in guiding sequential tasks.

The Occlusion Technique

The issue of the division of visual attention also raises important questions about the usefulness of the occlusion technique for predicting task completion times and the underlying theory of foveal visual sampling proposed by Wierwille (1993). The occlusion method is based on the idea that spatially separate visual information is sampled in a series of distinct glances, which, in a driving context, would be to the IVIS, then the road, followed by the IVIS, and so on (Wierwille, 1993). Shutter glasses are used to completely occlude the visual scene to replicate glances away from the road, toward the IVIS. If the hypothesis proposed in the current study is correct (i.e., that the visual resource can be shared to some extent between focal and peripheral areas), then occluding the entire visual scene would not represent real glance behaviour. Although Pettitt et al. (2007) were able to make accurate predictions of task times (within 20% of measured times) using the occlusion theory, it may be the case that the underlying mechanisms of glance behaviour in a dual-task environment are not reflected in the occlusion model. Wierwille's work on visual sampling was conducted two decades ago, before the widespread introduction of screen-based IVIS. Although he reported the use of peripheral vision for monitoring the road scene (Wierwille, 1993), it is unlikely that this would have been proposed as a method for attending to secondary tasks because of the nature of traditional controls. Little peripheral processing would be expected to occur in relation to traditional hard dashboard controls and static displays, as peripheral vision is most sensitive to movements and changes in a dynamic scene (Wittmann et al., 2006). The introduction of dynamic, visual displays associated with more modern screen-based IVIS will have increased the potential for peripheral processing because displays are usually located in closer proximity to the road scene: therefore, the theory may need to be extended to include peripheral attention to the IVIS, rather than focussing on foveal sampling alone. It is clear that further work is needed to investigate how a more integrated modal processing approach can be built into existing models, including the extended CPA approach presented in this study, to enable more accurate predictions of IVIS task times.

Road and IVIS Glance Durations

Wierwille's (1993) visual sampling model and the theory underlying the occlusion technique may no longer accurately represent drivers' visual behaviour as they do not account for modern IVIS, which are located in closer proximity to the forward road view. This is one possible explanation for the shorter glance times observed in the current study and in recent work on dual-task driving performance. For example, Sodhi et al. (2002) reported average on- and off-road glance durations of 760 ms for off-road glances and 420 ms for on-road glances. Tsimhoni et al. (1999) used an occlusion interval of 500 ms to represent off-road glances, and Green and Tsimhoni (2001) suggested that natural road fixations are around 500 ms. These values are similar to the glance durations found in the current study and are likely to be more representative of typical glances to the IVIS and road than the ISO occlusion values, which represent *maximum* glance times. Glance durations are also likely to have been affected by the way in which information presented by the IVIS was chunked (Wierwille, 1993). No tasks required drivers to read large amounts of text or scroll through long lists: these tasks generally require longer periods of continuous attention. Tasks used in the current study could be interrupted and resumed at any point between individual target selections without affecting the status of the device: this may have enabled the relatively short glances at the IVIS that were found in the study.

The use of shared glances by drivers may also be a cause of the shorter glance durations observed in this study; for example, it is assumed that a driver fixates on a particular object (either the road, and particular features in the road environment, or the IVIS), but can also use peripheral vision to monitor an area that is spatially separate from the point of fixation. A 'spotlight' metaphor has been used to describe how a single 'beam' of attention highlights information that falls within a specific spatial area (see Broadbent, 1982). More recent neuropsychology research has indicated that this spotlight can be divided between spatially separate locations (e.g., Driver and Bayliss, 1989; Muller et al., 2003), and this evidence supports the hypothesis that targets in the foveal and peripheral regions of vision can be attended to simultaneously. Studies of visual attention in driving have also differentiated between attention in foveal and peripheral vision (e.g., Crundall et al., 2002; Summala et al., 1996; Mourant and Rockwell, 1970), and have found that successful monitoring of the road whilst the fovea is fixed on the IVIS can be achieved to some extent via the peripheral visual resource (Mourant and Rockwell, 1970). If the driver is able to successfully monitor the road and IVIS simultaneously during shared glances, then more information about both targets can be obtained, reducing the need for longer glances at the IVIS or the road in isolation. Use of peripheral vision for monitoring information is, however, dependent on the area covered by a person's glance, and this can be influenced by a number of factors. Previous studies have observed narrowing of the visual field caused by increased workload, that is, as imposed by a complex driving scenario (Crundall et al., 2002; Rantanen and Goldberg, 1999; Summala et al., 1996). A condition of 'tunnel vision' can be produced by certain conditions, in which participants have been instructed to focus on a high-demand foveal task and there is pressure to complete the task (Crundall et al., 2002). This tunnel vision involves a general decrease in peripheral vision (Rantanen and Goldberg, 1999). There are also limits to the type of information that can be perceived via peripheral vision (Flannagan and Sivak, 1993); for example, studies have shown that peripheral vision is not adequate for successful hazard detection (Summala et al., 1996; Harvey et al., 2011c). This is supported by research in neuropsychology that has found that although people are capable of attending to two spatially separate stimuli via two attention 'spotlights', the perceptual representations are limited in resolution in comparison to a singularly attended object (McMains and Somers, 2004). In other words, people have difficulty *detecting* different targets even if they are capable of monitoring two separate sources without interference (Pashler, 1998; Posner and Petersen, 1990), and hazard detection may be impaired as a consequence. It is therefore likely that in high-workload conditions or in hazardous driving environments, drivers may not be able to employ shared glances and would have to rely on a pattern of single glances, alternating between the IVIS and the road, that may be more similar to the behaviour proposed by Wierwille (1993). The visual sampling strategy employed by drivers is therefore likely to be influenced by conditions in the driving environment and by drivers' perceptions of the risks posed by the hazards and events in this environment. The ability to detect particular events or information in the visual scene via shared glances and the consequences of this for drivers' visual sampling strategies will be an important area for future investigation.

IMPLICATIONS FOR IVIS DESIGN

The findings presented in this chapter have important implications for the design of IVIS; for example, a driver's ability to share visual attention is dependent on the characteristics of the task, including operation types and familiarity to users (Ho and Spence, 2008). It is also dependent on the location of the IVIS display screen in relation to the driver's UFOV (Rogé et al., 2002). Visual distraction from the road could be reduced by designing tasks that are suited to pre-attentive processing and locating the IVIS display within the driver's UFOV to allow peripheral processing to take place, reducing task times (Mourant et al., 1980; Wittmann et al., 2006; Burnett, 2000). Systems that use remote input usually have a display which can be located in the driver's line of sight, and this may offer an advantage over direct-touch systems, such as the touch screen, which must have a display screen located within the driver's zone of reach and may consequently be further from the UFOV.

LIMITATIONS OF THE CPA MODEL

The division of visual attention and processing between focal and peripheral vision is not well understood (Haslegrave, 1993; Snowden et al., 2006; Sodhi et al.; 2002; Wierwille, 1993). Further work is needed to investigate the extent of pre-attentive processing and peripheral attention during driving and for different IVIS tasks. Greater understanding of the type and amount of information that can be processed by peripheral vision, and the factors that influence this, will allow more accurate modelling of task times across a wide range of IVIS tasks. Visual field and ability to detect targets in the periphery also varies between individual drivers and different task scenarios (Rogé et al., 2002; Sodhi et al., 2002). The model will therefore always produce some level of inaccuracy in task time predictions. With age, visual processing slows down (Collet et al., 2010b), accommodative ability declines (Lockhart and Shi, 2010; Olson, 1993), and there is evidence that peripheral detection in a dual-task environment degrades (Rogé et al., 2004). This is likely to cancel out any reduction in task time due to pre-attentive processing for older drivers. Monotonous, prolonged driving also reduces a driver's ability to process peripheral visual information, as vigilance decreases (Rogé et al., 2004). Factors identified here as possible causes of prediction error (age, familiarity, driving conditions, task type) should be considered for inclusion in future iterations of the model to allow predictions of task time to be made across a wider range of user groups and driving environments (Young and Stanton, 2002).

Whilst vision is considered the most important processing mode in the dual-task driving environment (Wierwille, 1993; Sivak, 1996), the assumption of a relationship between eye movements and attention (Sodhi et al., 2002) is not correct in all cases (Shinar, 2008; Kramer and McCarley, 2003; Hoffman, 1998). This can be demonstrated in the case of inattention blindness. Inattention blindness describes the situation in which a driver's gaze is fixated on a target but the driver does not attend to the target and therefore does not perceive it (Goldstein, 2002; Strayer and Drews, 2007; Senders et al., 1967). In other words, the driver 'looked but did not see' (Herslund and Jørgensen, 2003; Langham et al., 2002; Shinar, 2008). This is evidence that it may not always be appropriate to rely on visual behaviour data for the prediction of attention to different tasks: attention can shift independently of the eyes; however, eye movements require visual attention to orient to a new target (Hoffman, 1998). It is also well known that humans do not attend to all of the available visual information in the environment (Schiffman, 2001; Kramer and McCarley, 2003), and therefore it seems that a more holistic approach to the evaluation of attention is needed (Kadar and Effken, 2005; Kramer and McCarley, 2003; Sivak, 1996), in which the interdependence of the visual, cognitive, auditory, and manual tactile processing modes is examined further. This is what, to some extent at least, has been achieved in this paper: the relationship between modal processes has been investigated to determine order and dependencies. CPA works on the premise that a decision needs to be made before a physical action can be started and that a target on a menu screen cannot be selected until the target to open that menu screen has been activated; however, what is still unclear is the extent of overlap between parallel processes (Wickens, 2002). This will be an important area for future investigation.

CONCLUSIONS

The aim of this study was to investigate whether an analytical model for the prediction of stationary IVIS task times could be extended to predict IVIS interaction times in a driving context with the addition of a model of drivers' visual behaviour. This was done for fastperson and slowperson estimates using 10th and 90th percentile data, respectively, as well as for median (middleperson) predictions. Two models of visual behaviour were proposed, based on the glance data collected in empirical tests of IVIS interaction in a dual-task driving environment. The shared glance CPA model, which categorised glances as IVIS-only, road-only, or shared, produced the most precise predictions of task time, compared to the empirical results. Mean prediction errors were within acceptable limits for the middleperson and slowperson predictions; however, there were instances of individual task time predictions that were outside the limits defined for each model version and further work is needed to explore the causes of these errors. Increasing the size of the sample from which the data were taken is likely to improve the precision of the models and will provide more information about the nature of IVIS interaction to enable more accurate modelling of individual operations and IVIS/on-road glances. The models could also be improved by integrating factors such as driver characteristics and environmental effects to produce predictions for a wider range of drivers and scenarios.

In developing and refining the CPA model for dual-task IVIS interaction, interesting visual behaviour patterns were identified. Examination of video data and gaze profiles from the empirical study appeared to show that visual time-sharing behaviour does not consist of simple, sequential glances to spatially separate targets, that is, IVIS-road-IVIS-, etc.; rather, shared glances in which both the road scene and LCD scene are attended to in parallel appear to be utilised in certain circumstances. This hypothesis is supported by findings reported in both the driving performance and neuropsychology literature that have shown that the 'spotlight' of attention can be split between two spatially separate objects. The extent and effectiveness of this shared visual attention is likely to vary according to characteristics of the task, user, and environment. The specific influence of these factors requires further investigation. The existence of a peripheral processing mechanism in the dual-task driving scenario affected how IVIS tasks were modelled and the CPA model has been developed to integrate shared glances; however, the effects of splitting attention on visual search times for targets in the IVIS need further investigation in order to adapt the models accordingly. The nature of CPA, which breaks down tasks into their smallest component operations and enables modelling of parallel processes, was shown to be suitable for analysing the division of attention at a detailed level. Previous theories and models of visual attention (e.g., Wierwille, 1993), including the occlusion technique (International Organization for Standardization, 2007), are based on the assumption that the visual time-sharing process can be modelled simply; however, the findings from this study indicate that visual attention is more complex to model accurately. Models of visual behaviour may also need to be updated to account for recent developments in IVIS interface technology, that is, the introduction of dynamic displays. This needs to focus on better integration of the theories of perception and attention with visual behaviour in a driving context.

9 Summary of Contributions and Future Challenges

INTRODUCTION

This book began by documenting the development of in-vehicle information, from the first speedometer to the touch screens and rotary dials of contemporary vehicles. From the research conducted as part of this work, it is clear that Ergonomics has always had a significant role to play in the development of the motor vehicle and with the proliferation of information and functionality available to the twenty-first century driver, Ergonomics now more than ever has the potential to exert significant positive influence on the safety, efficiency, and enjoyment of driving. In light of the recent developments in in-vehicle interaction technologies, namely, the shift from hard dashboard controls to digital screen-based systems, the main aim of the work presented in this book was to investigate how the usability of these technologies can be modelled and evaluated. In realising this aim, it was essential to take account of the unique dual-task scenario produced by driving and simultaneously interacting with secondary in-vehicle tasks. The main findings of the work are summarised below, followed by a discussion of the implications of the research based around some key questions that arose during the development of the book. Finally, key areas for future research are presented.

SUMMARY OF THE FINDINGS

The work presented in this book was structured around three key objectives as outlined in Chapter 1: the findings are summarised in relation to these objectives.

The first objective was to define and understand usability in the context of IVIS. A review of the literature highlighted the recent increase in the importance of the concept of usability, which was originally introduced in a purely HCI context, but today is considered to be a central goal of product design. Many references to usability were found in the literature; however, definitions of the concept were disparate and in numerous cases nonexistent. Although existing definitions captured general principles, including effectiveness, efficiency, and satisfaction (e.g., ISO, 1998), they were inadequate for identifying more detailed criteria, leaving the concept of usability open to interpretation. A common theme that was drawn from the literature review was the importance of context-of-use in defining usability criteria. This stemmed from the emergence of the idea that usability is not an intrinsic function of the product; rather, it is determined by the characteristics of the users, tasks, and system environment. To address this issue, a context-specific definition of usability was developed for IVIS. The existing definitions of usability were used to guide the identification of specific criteria that were most relevant in an IVIS context.

The second objective was to develop a multimethod framework to support designers in the evaluation of IVIS usability. The aim of creating the context-specific definition of usability was to enable the interaction between the user, tasks, and system to be modelled and evaluated. A framework was developed to structure the modelling and evaluation process. The main body of work has focussed on applying this framework in a series of case studies in order to validate and, where appropriate, extend the evaluation methodology for use in an IVIS context. The framework can be split into two sections: analytic methods and empirical methods. These methods were used to evaluate existing IVIS and the results served two major purposes. First, the utility of the methods was investigated in order to explore specific application areas in the context of IVIS and to identify targets for further development of evaluation techniques. The analytic and empirical evaluation case studies were approached with an aim to enhance the IVIS design process, rather than with the intention to make absolute measurements of system performance. Second, the evaluation methods enabled comparisons to be made between different IVIS input types, based on characteristics of the devices and their influences on primary driving performance. A distinction was made between direct and indirect input device types, following Rogers et al. (2005). The findings of the analytic and empirical evaluation case studies indicated that the suitability of the input type is largely dependent on task type and design of the GUI. This signified a need for a product development process that integrates the design of all interface components (input device, task structure, GUI design) in an iterative process, to produce an IVIS which is optimised for best performance. These findings also identified a need for a multimodal solution to driver–IVIS interaction, in which multiple input types are available to the driver to support the wide variety of task types offered by the system. The identification of usability issues, which can be used to inform the design of IVIS, was demonstrated in the empirical evaluation case study. The findings of the analytic methods case study highlighted a trade-off between the objectivity of a method and focus on context-of-use. A need for an objective method for early Ergonomics analysis, which integrated context-of-use factors for the dual task driving scenario, was identified: this drove the project towards the development of the CPA technique for predictions of dual-task IVIS interaction times.

The third objective was to develop an analytic usability evaluation method that enables useful predictions of task interaction, whilst accounting for the specific context-of-use of IVIS. The CPA method was selected for inclusion in the evaluation framework because it is capable of modelling parallel operations in a task: this made it particularly suitable for extension to the dual-task driving scenario. Initial CPA task time predictions enabled comparisons between IVIS; however, the lack of precision in these estimates highlighted the need for a more context-specific consideration of individual operation timings within the CPA model. Empirical studies of single- and dual-task IVIS performance provided the information and timing data to improve the precision of the CPA and enabled the expansion of the model for fastperson and slowperson predictions, as well as estimates of median (middleperson) task times. To create the dual-task model of driver–IVIS interaction, a model of visual behaviour based on the glance switching pattern proposed by Wierwille (1993) was integrated into the CPA. Analysis of the visual behaviour recorded in the empirical studies of driver–IVIS interaction showed that, rather than vision being a single resource that is split into separate glances to the road scene, followed by the IVIS, and so on, drivers employ 'shared glances' in which information from different sources is processed simultaneously. A pattern of visual behaviour that integrated these shared glances was incorporated into the CPA model and produced task time predictions with acceptable precision for the majority of the modelled tasks.

Novel Contributions of the Work

The novel contributions of this work can be summarised as follows.

Definition and Further Understanding of Usability in the Context of IVIS

A definition of IVIS usability, in the form of 6 context-specific factors and 12 measurable criteria, has been developed as part of this work. The context-of-use factors most important for driver–IVIS interactions were defined as dual-task environment, environmental conditions, range of users, training provision, frequency of use, and uptake. Elements from general definitions of usability were linked to the contextual factors to produce a set of measurable criteria, against which IVIS usability can be assessed. This was essential as a foundation for usability evaluations that accounted for the context-of-use of IVIS and the needs of the driver.

Development of a Multimethod Evaluation Framework to Support Designers and Evaluators of IVIS throughout the Product Development Life Cycle

A framework consisting of analytic and empirical methods was developed to guide IVIS designers through a comprehensive usability evaluation, starting at the very earliest stages in product development. Analytic methods were included in the framework to enable early stage predictions of task times and error rates and qualitative assessments of usability and GUI design. Empirical methods can be applied at a later stage of product development to measure users' interactions with an IVIS and the effect of this on driving. The combination of methods enables a comprehensive evaluation of all aspects of IVIS usability.

Demonstration, via Case Studies, of an Effective Usability Evaluation Consisting of Analytic and Empirical Methods

The information inputs and outputs, application stage, and resources were explored for each of the methods in the framework. The framework's multimethod approach enables a comprehensive assessment of usability, producing performance predictions and absolute measurements that can be used to identify usability issues and analyse IVIS performance and its effect on driving. Estimates of training and application times will also be useful for future users of the framework. Methods that are quicker to apply, such as heuristic analysis, tend to provide a more qualitative overview of system usability, whereas methods that have higher time and resource demands, such as CPA and measures of actual user performance, tend to provide more quantitative results relating to more specific aspects of usability.

Comparative Evaluation of Automotive Manufacturers' Existing Driver–Vehicle Interaction Strategies

The framework was applied in evaluations of some of the most commonly used IVIS interfaces: this has contributed to our understanding of the factors that are important in existing interaction strategies. Indirect input devices increase task times because the actions of the driver have to be translated into movements on screen and the driver must monitor and understand this translation in order to perform the task successfully. Usability problems will occur if the structure of tasks and the design of the GUI are not optimised for the particular input device being used. This knowledge can be used to inform future design decisions and will be particularly important in the development of multimodal interaction strategies.

Extension of the CPA Method for Fastperson and Slowperson Predictions of IVIS Task Time for Estimates of Best- and Worst-Case Scenario Performance

In order to address one of the most important principles in Ergonomics, the CPA model was extended to produce fastperson and slowperson predictions. These represent the lower and upper bounds of performance, encouraging designers to account for variation over a wide range of the user population, rather than focussing only on the 'average' user. Slowperson predictions will be useful for comparisons against maximum recommended task times, such as that specified by the 15-second rule. Fastperson and slowperson predictions might also be used to estimate the effect of particular user characteristics on IVIS interactions: for example, fastperson predictions are likely to approximate to an experienced user's performance and slowperson predictions to an older driver's performance.

Extension of the CPA Method for Application in a Dual-Task Driving Environment, Integrating a Model of Visual Behaviour for the Prediction of IVIS Task Times

A pattern of IVIS-only, road-only, and shared glances was integrated into the CPA model to represent the division of attention between primary and secondary driving tasks. The visual mode is most widely used during driving and was therefore an important parameter to integrate into the model of dual-task performance. Glances at the road interrupted IVIS operations, and this increased the total time required for task completion. The model will be useful to designers for predicting the efficiency of IVIS interfaces under dual-task driving conditions.

Validation of CPA Predictions against Data Gathered in Empirical User Tests Using a Driving Simulation

In order to assess the precision of the CPA model, the predicted task times were compared to actual task times recorded in a simulated driving environment with a sample of users. The empirical study was used to measure task interaction times with a touch screen IVIS, similar to one used by a major automotive manufacturer. The model predictions were compared against empirical data for 10th, 50th, and 90th percentile drivers. The CPA model produced accurate mean task time predictions for IVIS interactions performed at the same time as driving. This validated the model for predictions of IVIS task times for young, novice IVIS users.

Development of a Hypothesis of Shared Glances in Visual Information Processing in a Dual-Task Driving Environment

Analysis of visual behaviour data collected in an empirical study of driver–IVIS interaction in a dual-task scenario indicated that drivers may use shared glances to obtain information from the IVIS and road scene simultaneously. This is likely to be a consequence of the introduction of dynamic IVIS displays in the vehicle and will be dependent on the proximity of visual information from the road and IVIS. This shared glance hypothesis has important implications for visual behaviour theories that model distinct glances at separate visual targets. It will also influence how visual behaviour is modelled and simulated: for example, the occlusion technique is based on a theory of separate road and IVIS glances, which may not actually be representative of dual-task visual sampling. Models and theories of visual behaviour may need to be updated to account for the shared glance hypothesis.

KEY QUESTIONS

The findings and their implications are further explored using a number of key questions, which arose during the course of the research:

WHAT HAS THE RESEARCH TOLD US ABOUT IVIS?

A technical specification for the 'perfect' IVIS was not the intended outcome of this work; rather, a deeper understanding of the nature of IVIS interaction (context-of-use and driver needs, Chapters 2 and 3) and a considered appraisal of methods for exploring usability (evaluation framework, Chapter 4) were the drivers for further research. The knowledge generated by these initial investigations provided the foundations for an exploration of the characteristics of IVIS interfaces that give rise to potential usability issues (analytic and empirical evaluation, Chapters 5 and 6). This information is intended to inform the design process for any IVIS and is underpinned by findings related to the distinction between direct and indirect input devices, the interaction of input device, GUI and task structure, and the need for multimodality in user–IVIS interactions. Furthermore, modelling the task–user–system interaction led to a focus on the investigation of modes of interaction and, particularly given the context, a focus on visual processing as the dominant mode in driving and secondary task performance (CPA model of IVIS interactions and visual behaviour, Chapters 7 and 8). This led to the extension of the MPCA method for modelling dual-task IVIS interactions. The theories of visual behaviour in driving have consequently been questioned in relation to the dynamic visual displays common in IVIS today. The findings will influence the design of visual IVIS information, positioning of displays within the vehicle and also the way in which driver–IVIS interactions are modelled in the future.

What Are the Requirements for a Successful Usability Evaluation?

One of the key messages of this research is the importance of the context within which a product or system is used (Chapter 2). The findings of this project have reinforced Bevan's (2001) notion that a product or system has a capability to be used in a particular context, rather than an intrinsic usability. The complex interaction between a system, its tasks, its users, and its environment must be understood in order to define this context-of-use (Chapter 3). Context determines how usability is defined, the criteria against which it is measured, and the methods which are most suited to its evaluation (Chapter 4). The framework was designed to encourage an iterative process of evaluation and redesign from an early stage in the product development process. The motivation for this was to increase the impact of Ergonomics at the beginning of the product life cycle in order to guide design in a proactive manner, rather than reacting to poor design when it is too late to make any significant improvements (Nowakowski et al., 2000; Pettitt et al., 2007; Stanton and Young, 1999a). An important conclusion of this work is that the aim of an evaluation methodology should not only be to measure one or more aspects of product usability, but also, more importantly, to inform the redesign of the product based on usability criteria and thorough consideration of context.

What Are the Advantages of the Evaluation Framework in an Industrial Context?

A need for early stage usability evaluation was one the underlying motivations for this work. In order to stay competitive, and with the important aim of reducing driver distraction, automotive manufacturers need to develop novel concepts to facilitate improved driver–IVIS interactions. Early stage, conceptual designs may be presented as paper-based diagrams or basic system specifications: the analytic methods in the framework were included to provide a useful evaluation of these concepts, as these methods generally have low system information requirements. The case study findings highlighted a trade-off between the objectivity of a method and the extent to which it accounted for context-of-use. Work with the industrial sponsor of this research also reinforced the need for a method that could produce an objective and quantitative measure of usability, whilst being quick and easy to apply at an early stage of product development. With these requirements and limitations in mind, the CPA method was extended for the prediction of dual-task IVIS interaction times. This also involved the development of the CPA calculator, with the aim of reducing the time taken to analyse tasks and to allow users to instantly see the effects of using different parameters in the model. This book aims to provide a simple and accessible resource for knowledge transfer from academia to industry and will encourage automotive manufacturers to apply HF methods with the aim of improving IVIS usability.

Summary of Contributions and Future Challenges

WHAT ARE THE ADVANTAGES OF THE CPA MODELLING METHOD?

CPA was initially selected over other task time prediction methods due to its relative simplicity and ability to model parallel processing modes (Baber and Mellor, 2001; Olson and Moran, 1996; Stanton and Baber, 2008). A range of analytic and empirical methods were investigated in case studies of IVIS evaluation. A trade-off between a method's objectivity and consideration of context-of-use was discovered, and this prompted an aim for the development of an evaluation technique that would produce quantifiable predictions of usability at an early stage in product development, whilst accounting for important contextual factors, particularly relating to the dual-task driving environment. CPA was selected for further development because it enabled quantifiable predictions of an important aspect of usability (task times) and its multimodal structure offered potential for the integration of primary driving behaviour parameters. Development of the CPA in this project took advantage of the method's multimodal approach by further partitioning the visual mode into separate sections based on glance target. The effect of the newly integrated visual behaviour pattern on parallel and serial operations and on the complete task could clearly be seen in the CPA diagrams, which, unlike other methods, clearly show the sequence and dependencies of operations in a task. This was further enhanced by the development of the CPA calculator (Chapter 7), which enables users to instantly see the effects of using different parameters in the model. The calculator dramatically reduced the time required to create CPA diagrams and calculate task times: this is expected to provide a real benefit to automotive manufacturers in early stage product evaluations.

A significant contribution of the CPA development work was to extend the predictions to fastperson and slowperson ranges, as well as median (middleperson) values (Chapter 7). An important principle in Ergonomics is that products and systems should be designed to accommodate a wide range of users (Dul and Weerdmeester, 2001). Focussing only on the average user excludes a large proportion of the target population; rather, IVIS designers should be interested in the range of possible interaction times, from best-to-worst performance. This will provide much more meaningful information regarding which tasks are suitable for operation whilst driving, the maximum level of distraction that could be caused by IVIS interaction, and the extent of the effects of the driver population's characteristics on IVIS performance.

WHAT ARE THE IMPLICATIONS OF THE CPA FINDINGS FOR VISUAL ATTENTION THEORY?

The findings of this research led to the development of a hypothesis of shared glances, which are used to obtain visual information from two separate sources simultaneously (Chapter 8). This was proposed as a more realistic model of visual behaviour in a dual-task environment than the time-sharing model proposed by Wierwille (1993). This shared glance hypothesis also has implications for the use of the occlusion technique for simulating the division of visual attention during driving. A previous criticism of the occlusion technique is its failure to accurately represent the effects of workload: this is because during occluded periods there is no workload imposed on the driver, whereas in reality the driver would be gathering and processing information during a glance away from the target scene (Gelau and Schindhelm, 2010; Monk and Kidd, 2007). The current research adds a further dimension to this problem as the occlusion technique fails to account for the visual information obtained during shared glances.

The hypothesis of shared glances also has implications for the design of visual IVIS information. This may offer an opportunity for the increased use of information that can be perceived successfully using peripheral vision during a shared glance. This would reduce eyes-off-road time if more information could be perceived at the same time as monitoring the driving task; however, the extent to which a driver's perception of the road scene is affected by sharing visual attention between two sources is not known. For example, does a driver perceive the road scene with the same level of detail and precision during a shared glance compared to during a road-only glance? Currently, there is little evidence to answer this question. Studies have suggested that successful monitoring of the road can be achieved to some extent via the peripheral visual resource (Mourant et al., 1980). This has been found to apply to lane keeping and other vehicle control activities; however, research has shown that peripheral vision in not adequate for successful hazard detection (Summala et al., 1996). Consequently, encouraging shared glances via presentation of specially designed IVIS information may not offer an advantage; rather, it may degrade a driver's response to hazards in the road environment, which will have an obvious negative influence on safety.

Further, these findings could have consequences for the positioning of in-vehicle displays. Locating the display as close as possible to the driver's line of sight is generally recommended on the premise that glances at the display will be of smaller magnitude and therefore take less time. Displays in close proximity to the driver's forward view will increase the ease with which drivers can make shared glances; however, if this behaviour leads to degradation of hazard detection, the advice should surely be to locate the display further from the field of view to discourage shared glances. Furthermore, if shared glances had only positive consequences for successful dual-tasking, head-up displays (HUD) would be an ideal IVIS solution; yet there is much evidence to show that this is not the case.

AREAS FOR FUTURE RESEARCH

EXTENDED APPLICATION OF THE EVALUATION FRAMEWORK AND CPA MODEL

The evaluation framework is aimed at all IVIS technologies, from concepts to fully integrated in-vehicle devices. In this project, the framework has been used to evaluate existing IVIS input devices, including a touch screen, rotary dial, and remote joystick controller. Work with the industrial sponsor of the research presented in this book has highlighted a requirement for novel IVIS concepts that can offer an improved level of usability, compared with the existing systems. Application of the evaluation methods at the concept design stage would also ensure a proactive Ergonomics approach to product development. The challenges of concept evaluation are related to limited information about the interaction style and structure of tasks and the lack of a sophisticated prototype system with which to simulate the user–system interaction. The analytic methods in the framework are suitable for resource-limited evaluations

Summary of Contributions and Future Challenges

as they generally have low information requirements. Future work should focus on the validation of these methods with concept technologies.

The CPA model was developed specifically around the touch screen, which is one of the most common IVIS interfaces in use by automotive manufacturers today. Much of the operation time data and HCI theory will also be applicable to alternative IVIS input devices; however, there will be some interaction styles that are unique to other existing or concept IVIS technologies. This might include visual processing of information displayed on a HUD, verbal input to a voice recognition system, and manual rotation of a rotary dial for menu navigation. Methods for estimating and measuring these operation times need to be explored in order to create an accurate database of CPA times that is applicable to any existing or concept IVIS.

There is also considerable potential for the evaluation framework to be extended for application to other domains, particularly other areas of transport, which will be subject to many of the same contextual factors, such as safety and multitasking, as road vehicles. The control of vehicles in domains such as rail, maritime, and aviation involves user–device interactions similar to those in road vehicles, and the methods outlined in the evaluation framework should be applicable in these scenarios. Many of these transport domains have not been as well researched as car driving in terms of the HMI and the influence of this on primary task performance. As with road vehicles, there are significant safety implications associated with the performance of human operators in rail, aviation, and maritime domains, and the design of the HMI will have a significant impact on this. There is therefore a serious need for greater understanding of how these interfaces can be designed for increased usability. A starting point should be the exploration of context-of-use for different domains: user characteristics, training, frequency of use, and environmental factors are all likely to differ between rail, aviation, and maritime applications, and there will be additional usability criteria that will need to be considered. The CPA should be easy to apply to these other areas as many of the basic operations will overlap. Work will be required to extend the database of operation times to incorporate more context-specific data: a future goal would be a transport-wide database of reference values for operation times, which would allow quick and easy application of CPA to any domain. The CPA model will also need to be validated against empirical data in any area of application.

EVALUATION OF IVIS USABILITY ACROSS ALL USER GROUPS

The empirical methods case study identified some important factors that affect the usability of IVIS. These included whether the IVIS interaction was performed in a stationary or moving vehicle and the optimisation between GUI design, structure of tasks, and input device. In the case studies presented in this book, the influence of these factors on usability was analysed for a sample of young, inexperienced IVIS users. It is likely that these factors may have different effects for different user groups and design decisions regarding input type and GUI should be based on evidence from the full range of potential users. Future work should aim to identify the combinations of input type, GUI design, and menu structure most suited to particular task types and user groups. It may be that a multimodal system, which offers users a choice of interaction styles, will provide an optimal and inclusive IVIS solution.

EXTENSION AND VALIDATION OF THE CPA OPERATION TIMES DATABASE

A limitation of the CPA models developed as part of this research was the small sample sizes from which the operation time data was drawn and against which the task times were compared. Future work must include the expansion and validation of the operation times database using larger samples. The aim would be to achieve sampling on the scale of anthropometric data collection (i.e., thousands of participants): this would be a significant improvement on current HCI data, such as that reported by Card et al. (1983), which was based largely on small samples of less than one hundred.

Larger user samples should also incorporate a wider range of user characteristics, particularly accounting for variation in age and experience. These factors would enable much more realistic estimates of fastperson and slowperson task times, which would have real relevance in the design of products and systems for the full range of potential users. Furthermore, user characteristics may not only influence operation times, but also multimodal processing and therefore the CPA structure. This needs to be investigated with the aim of establishing rules for how individual differences shape the task–user–system interaction.

INVESTIGATION OF FOCAL AND PERIPHERAL VISION IN A DUAL-TASK ENVIRONMENT

Future work should aim to investigate the ways in which modern dynamic IVIS displays alter the assumptions of traditional models of visual sampling behaviour (e.g., Wierwille, 1993). The hypothesis of shared glances, presented in this research, should be considered as an alternative method of processing, which has arisen due to the changing nature of IVIS display presentation. In order to investigate the hypothesis, more needs to be known about the extent of visual processing in the focal and peripheral fields. This will need to be explored in empirical tests designed to measure drivers' visual processing in a dual-task environment. A number of independent variables will be of interest, including the proximity of the visual display to the forward road view, the type of IVIS task and visual information available, and user characteristics such as age that are likely to affect peripheral processing. Dependent variables will include secondary task performance to measure the effects of shared glances on IVIS interaction. Visual perception of the road scene and the consequences for driving performance during shared glances will also need to be investigated in order to make design decisions about the facilitation of shared glances. An aim would be the development of a set of rules to determine the types of glance used to obtain different visual information, the effects of the proximity of information on use of peripheral vision, and possible detrimental effects on safe driving: this would be applicable at an operation level and could therefore be built into the CPA model.

Development of the CPA model was based on an assumption that the nature of task interaction did not change from a static task environment to a dynamic, dual-task environment. It is likely, however, that interruptions to the secondary task, caused by

Summary of Contributions and Future Challenges 185

attention to the road scene, would alter the way in which drivers resumed the IVIS interaction (Monk and Kidd, 2007; Noy et al., 2004). For example, they may need to reacquire information lost in the glance away from the IVIS (Nowakowski et al., 2000). This will also need to be investigated in tests of the shared glance model.

CONCLUDING REMARKS

The goal of this work has been to improve the product development and evaluation process to enable the production of more usable IVIS interfaces. During this process, we discovered disparate definitions of usability and an even less unified approach to usability evaluation in both academia and industry. The realisation of the importance of context-of-use meant that the failure to achieve a universal definition of usability was unsurprising; however, this does not mean that comprehensive usability evaluation should not be attempted. Experience with industry during this research project also showed that usability evaluation had to be made more accessible and useful to those applying the methods and this has been an aim for the development of the evaluation framework. Usability evaluation methods have been critically analysed, and many shortcomings were identified; however, the CPA development work has demonstrated that knowledge of the context-of-use can be used effectively to enhance IVIS usability evaluation.

This work has not just been about how usability can be measured but also what usability actually means in the context of IVIS. As we discussed in the introduction to this book, the concept of usability has been present in the vehicle design world in some form from the very early days of the automobile; however, the Ergonomics challenges of today are very different from a century ago and the car continues to evolve to meet the needs of the contemporary driver in an increasingly information rich world. The scope of Vehicle Ergonomics has expanded to encompass not only physical comfort issues such as seat design and reach but also cognitive factors relating to information processing, workload, and situational awareness. This has been motivated by an increasing focus on the driver's capabilities and a continuing need to improve the safety and efficiency of the vehicle whilst maintaining the enjoyment that has always been associated with the freedom and ease of mobility that car driving provides. A major development has been the introduction of secondary functions into the vehicle, such as entertainment, comfort, navigation, and communication. This has changed the nature of driving, giving the driver many more tasks to control and monitor on top of the primary driving duties of hazard detection and avoidance, vehicle control, and route guidance. Furthermore, the evolution of in-vehicle information presentation is expected to continue and, whilst we expect further increases in the amount and variety of in-vehicle functions over the coming years, there is also the likelihood of increasing automation of vehicle tasks, which has the potential to dramatically shift the role of the driver from active operator to passive monitor. The role of the ergonomist and vehicle designer is to not only keep up with the evolution of vehicle design but to anticipate future technological developments and driver needs. Questions for the Ergonomics discipline and automotive industry to address will centre around issues such as multimodal interaction, which is a likely next step for in-vehicle interface design, with voice recognition and haptic feedback becoming increasingly common and reliable technologies in vehicles. However, more needs to be known about the demands of multimodal processing in a driving environment. On the other hand, there is currently much focus on the Ergonomics issues of automation in many domains including driving. This is a development which is likely to have the opposite effect to multimodal processing: the issue will be not how to manage an increase in processing demands but how to cope with a reduction in this demand on the driver when tasks are shifted from driver-controlled to vehicle-controlled. The methods and approaches described in this book can and should be used to investigate and predict the likely effects of these developments on driver–vehicle interactions. Early stage evaluation will enable the automotive industry and researchers to anticipate the impact of the evolving vehicle on drivers, and we are optimistic that this knowledge will lead designers naturally toward a more Ergonomic approach to IVIS development and the design of the driver–vehicle interaction in general. We also hope that the ideas that have emerged as a result of the modelling work presented in this book are taken forward by other researchers to further the understanding of visual processing in the dual-task driving environment.

References

Abraham, S., 1979. Weight and height of adults 18–74 years of age. United States vital health statistics data from the National Health Survey, series 11. National Centre for Health Statistics, Rockville, MD.

Ackerman, P.L., Cianciolo, A.T., 1999. Psychomotor abilities via touch-panel testing: Measurement innovations, construct, and criterion validity. *Human Performance*. 12, 231–273.

Adler, P.S., Winograd, T.A., 1992. *Usability: Turning technologies into tools*. University Press, Oxford.

Agah, A., 2000. Human interactions with intelligent systems: Research taxonomy. *Computers and Electrical Engineering*. 27, 71–107.

Alliance of Automobile Manufacturers, 2006. Statement of principles, criteria and verification procedures on driver interactions with advanced in-vehicle information and communication systems. AAM, Washington DC.

Alonso-Ríos, D., Vázquez-García, A., Mosqueira-Rey, E., Moret-Bonillo, V., 2010. Usability: A critical analysis and a taxonomy. *International Journal of Human-Computer Interaction*. 26, 53–74.

Amditis, A., Polychronopoulos, A., Andreone, L., Bekiaris, E., 2006. Communication and interaction strategies in automotive adaptive interfaces. *Cognition, Technology and Work*. 8, 193–199.

Anderson, J.R., Lebiere, C., 1998. *The atomic components of thought*. Lawrence Erlbaum, Hillsdale, NJ.

Andre, A.D., Wickens, C.D., October 1995. When users want what's not best for them. *Ergonomics in Design*, 10–13.

Angell, L., Auflick, J., Austria, P.A., Kochhar, D., Tijerina, L., Biever, W., Diptiman, T., Hogsett, J., Kiger, S., 2006. Driver workload metrics task 2 final report. Crash Avoidance Metrics Partnership, Farmington Hills, MI.

Annett, J., 2002. Subjective rating scales: Science or art? *Ergonomics*. 45, 966–987.

Anttila, V., Luoma, J., 2005. Surrogate in-vehicle information systems and driver behaviour in an urban environment: A field study on the effects of visual and cognitive load. *Transportation Research Part F: Traffic Psychology and Behaviour*. 8, 121–133.

AT&T, 2012. 1946: First mobile telephone call [Online]. Available from: http://www.corp.att.com/attlabs/reputation/timeline/46mobile.html. [Accessed 15.05.2012].

Au, F.T.W., Baker, S., Warren, I., Dobbie, G., 2008. Automated usability testing framework. *9th Australasian User Interface Conference*, Wollongong, Australia, January 22–25, 2008, Australian Computer Society.

Averbach, E., Coriell, A.S., 1961. Short-term memory in vision. *Bell System Technical Journal*. 40, 309–328.

Azuma, S., Nishida, K., Hori, S., 1994. The future of in-vehicle navigation systems. *Vehicle Navigation and Information Systems Conference*, Yokohama, Japan, August 31–September 2, 1994, IEEE.

Baber, C., 2002. Commentary: Subjective evaluation of usability. *Ergonomics*. 45, 1021–1025.

Baber, C., 2005a. Critical path analysis for multimodal activity, in: N. Stanton, A. Hedge, K. Brookhuis, E. Salas and H. Hendrick (eds.). *Handbook of human factors and ergonomics methods*. CRC Press, London, pp. 1–8.

Baber, C., 2005b. Evaluation in human-computer interaction, in: J.R. Wilson and N. Corlett (eds.). *Evaluation of human work*. Taylor & Francis, London, pp. 357–387.

Baber, C., Butler, M., 2012. Expertise in crime scene examination: Comparing search strategies of expert and novice crime scene examiners in simulated crime scenes. *Human Factors*. 54, 413–424.

Baber, C., Mellor, B., 2001. Using critical path analysis to model multimodal human-computer interaction. *International Journal of Human-Computer Studies*. 54, 613–636.

Baber, C., Stanton, N.A., 1996. Human error identification techniques applied to public technology: Predictions compared with observed use. *Applied Ergonomics*. 27, 119–131.

Bach, K.M., Jæger, M.G., Skov, M.B., Thomassen, N.G., 2008. You can touch but you can't look: Interacting with in-vehicle systems. *CHI 2008 Conference on Human Factors in Computing Systems*, Florence, 5–10 April, 2008, ACM.

Baldwin, C.L., 2002. Designing in-vehicle technologies for older drivers: Application of sensory-cognitive interaction theory. *Theoretical Issues in Ergonomics Science*. 3, 307–329.

Bangor, A., Kortum, P.T., Miller, J.T., 2008. An empirical evaluation of the system usability scale. *International Journal of Human-Computer Interaction*. 24, 574–594.

Barón, A., Green, P., 2006. Safety and usability of speech interfaces for in-vehicle tasks while driving: A brief literature review. University of Michigan Transportation Research Institute (UMTRI), Ann Arbor, MI.

Baumann, M., Keinath, A., Krems, J.F., Bengler, K., 2004. Evaluation of in-vehicle HMI using occlusion techniques: Experimental results and practical implications. *Applied Ergonomics*. 35, 197–205.

Bayly, M., Young, K.L., Regan, M.A., 2009. Sources of distraction inside the vehicle and their effects on driving performance, in: M.A. Regan, J.D. Lee and K.L. Young (eds.). *Driver distraction: Theory, effects and mitigation*. CRC Press, Boca Raton, FL, pp. 191–213.

Beirness, D.J., Simpson, A.P., Pak, A., 2002. The road safety monitor: Driver distraction. Traffic Injury Research Foundation, Ontario, Canada.

Bevan, N., 1991. What is usability? *4th International Conference on Human-Computer Interaction*, Stuttgart, September 1–6, 1991, Elsevier.

Bevan, N., 1995. Usability is quality of use. *6th International Conference on Human-Computer Interaction*, Tokyo, July 9–14, 1995, Elsevier.

Bevan, N., 1999. Design for usability. *8th International Conference on Human-Computer Interaction*, Munich, August 22–26, 1999, Lawrence Erlbaum.

Bevan, N., 2001. International standards for HCI and usability. *International Journal of Human-Computer Studies*. 55, 532–552.

Bevan, N., Macleod, M., 1994. Usability measurement in context. *Behaviour and Information Technology*. 13, 1994.

Bhise, V.D., Dowd, J.D., Smid, E., 2003. A comprehensive HMI evaluation process for automotive cockpit design. *SAE World Congress*, Detroit, MI, March 3–6, 2003, SAE International.

Birrell, S.A., Young, M.S., Jenkins, D.P., Stanton, N.A., 2011. Cognitive work analysis for safe and efficient driving. *Theoretical Issues In Ergonomics Science*. 13, 430–449.

Blandford, A., Rugg, G., 2002. A case study on integrating contextual information with analyltical usability evaluation. *International Journal of Human-Computer Studies*. 57, 75–99.

Booth, P., 1989. *An introduction to human-computer interaction*. Lawrence Erlbaum, East Sussex.

Bouchner, P., Novotný, S., Pieknίk, R., 2007. Objective methods for assessments of influence of IVIS (in-vehicle information systems) on safe driving. *4th International Driving Symposium on Human Factors in Driver Assessment, Training and Vehicle Design*, Stevenson, WA, July 9–12, 2007, Elsevier.

References

Broadbent, D.E., 1982. Task combination and selective intake of information. *Acta Psychologica*. 50, 199–326.

Brook-Carter, N., Stevens, A., Reed, N., Thompson, S., 2009. Technical note: Practical issues in the application of occlusion to measure visual demands imposed on drivers by in-vehicle tasks. *Ergonomics*. 52, 177–186.

Brooke, J., 1996. SUS: A 'quick and dirty' usability scale, in: P.W. Jordan, B. Thomas, B.A. Weerdmeester and I.L. Mcclelland (eds.). *Usability evaluation in industry*. Taylor & Francis, London, pp. 189–94.

Broström, R., Bengtsson, P., Axelsson, J., 2011. Correlation between safety assessments in the driver-car interaction design process. *Applied Ergonomics*. 42, 575–582.

Broström, R., Engström, J., Agnvall, A., Markkula, G., 2006. Towards the next generation intelligent driver information system (IDIS): The Volvo car interaction manager concept. Volvo Car Corporation, Gothenburg.

Buckle, P., 2011. The perfect is the enemy of the good—ergonomics research and practice. IEHF Annual Lecture, 2010. *Ergonomics*. 54, 1–11.

Bullinger, H.-J., Dangelmaier, M., 2003. Virtual prototyping and testing of in-vehicle interfaces. *Ergonomics*. 46, 41–51.

Burnett, G., 2000. Usable vehicle navigation systems: Are we there yet? *Vehicle Electronic Systems European Conference and Exhibition*, Leatherhead, UK, June 29–30, 2000, ERA Technology.

Burnett, G., Porter, J.M., 2001. Ubiquitous computing within cars: Designing controls for non-visual use. *International Journal of Human Computer Studies*. 55, 521–531.

Burnett, G., Summerskill, S.J., Porter, J.M., 2004. On-the-move destination entry for vehicle navigation systems: Unsafe by any means? *Behaviour and Information Technology*. 23, 265–272.

Burns, P., Harbluk, J., Foley, J.P., Angell, L., 2010. The importance of task duration and related measures in assessing the distraction potential of in-vehicle tasks. *Second International Conference on Automotive User Interface and Interactive Vehicular Applications—Automotive UI*, Pittsburgh, PA, November 11–12, 2010, ACM.

Burns, P.C., Trbovich, P.L., Harbluk, J.L., McCurdie, T., 2005. Evaluating one screen/one control multifunction devices in vehicles. *19th International Technical Conference on the Enhanced Safety of Vehicles*, Washington, DC, June 6–9, 2005, NHTSA.

Busswell, G.T., 1922. *Fundamental reading habits: A study of their development*. University of Chicago, Chicago, IL.

Butler, K.A., January 1996. Usability engineering turns 10. *Interactions*, 58–75.

Butters, L.M., Dixon, R.T., 1998. Ergonomics in consumer product evaluation: An evolving process. *Applied Ergonomics*. 29, 55–58.

Byrne, M.D., 2001. ACT-R/PM and menu selection: Applying a cognitive architecture to HCI. *International Journal of Human Computer Studies*. 55, 41.

Cacciabue, P.C., Martinetto, M., 2006. A user-centred approach for designing driver support systems: The case of collision avoidance. *Cognition, Technology and Work*. 8, 201–214.

Caple, D.C., 2010. The IEA contribution to the transition of ergonomics from research to practice. *Applied Ergonomics*. 41, 731–737.

Car Connectivity Consortium, 2011. Car Connectivity Consortium unveils MirrorLink Roadmap [Online]. Available from: http://www.terminalmode.org/news-and-events/press-releases.html. [Accessed 15.05.2012].

Card, S.K., English, W.K., Burr, B.J., 1978. Evaluation of mouse, rate-controlled isometric joystick, step keys, and text keys for text selection on a CRT. *Ergonomics*. 21, 601–613.

Card, S.K., Moran, T.P., Newell, A., 1983. *The psychology of human-computer interaction*. Lawrence Erlbaum, London, UK.

Carroll, J.M., 2010. Conceptualizing a possible discipline of human-computer interaction. *Interacting with Computers*. 22, 3–12.

Carsten, O.M.J., 2004. Implications of the first set of HASTE results on driver distraction. Behavioural research in road safety 2004: Fourteenth seminar, London, Department for Transport.

Carsten, O.M.J., Merat, N., Janssen, W.H., Johansson, E., Fowkes, M., Brookhuis, K.A., 2005. HASTE final report. Institute for Transport Studies, University of Leeds, Leeds.

Cellario, M., 2001. Human-centered intelligent vehicles: Toward multimodal interface integration. *IEEE Intelligent Transport Systems*. 16, 78–81.

Chamorro-Koc, M., Popovic, V., Emmison, M., 2008. Human experience and product usability: Principles to assist the design of user-product interactions. *Applied Ergonomics*. 40, 648–656.

Chapanis, A., 1950. *Research techniques in human engineering*. John Hopkins University Press, Baltimore, MD.

Cherri, C., Nodari, E., Toffetti, A., 2004. Review of existing tools and methods: AIDE deliverable 2.1.1. Volvo Technology Corporation (VTEC), Gothenburg.

Chestnut, J., Aryal, B., Gellatly, A.W., Kleiss, J.A., 2005. A user-based qualitative usability assessment and design support assessment tool. *SAE World Congress*, Detroit, MI, April 11–14, 2005, Society of Automotive Engineers.

Chiang, D.P., Brooks, A.M., Weir, D.H., 2004. On the highway measures of driver glance behavior with an example automobile navigation system. *Applied Ergonomics*. 35, 215–223.

Cobb, J.G., 2002. Menus behaving badly. *The New York Times*, [Online]. The New York Times Company. Available from: http://www.nytimes.com/2002/05/12/automobiles/menus-behaving-badly.html. [Accessed 21.06.2012].

Cockburn, A., Gutwin, C., Greenberg, S., 2007. A predictive model of menu performance. *CHI 2007 Conference on Human Factors in Computing Systems*, San Jose, CA, April 28–May 3, 2007, ACM.

Collet, C., Guillot, A., Petit, C., 2010a. Phoning while driving I: A review of epidemiological, psychological, behavioural and physiological studies. *Ergonomics*. 53, 589–601.

Collet, C., Guillot, A., Petit, C., 2010b. Phoning while driving II: A review of driving conditions influence. *Ergonomics*. 53, 602–616.

Commission of the European Communities, 2008. Commission recommendation of 26/V/2008 on safe and efficient in-vehicle information and communication systems: Update of the European statement of principles on human-machine interface. Commission of the European Communities, Brussels.

Cox, K., Walker, D., 1993. *User-interface design*. Prentice Hall, London.

Crundall, D., Underwood, G., Chapman, P., 2002. Attending to the peripheral world while driving. *Applied Cognitive Psychology*. 16, 459–475.

Cunningham, W., 2007. Driving it: Car interfaces and usability. *CNET Reviews* [Online]. CBS Interactive. Available from: http://reviews.cnet.com/4520-10895_7-6744922-1.html. [Accessed 21.06.2012].

Daimler, 2012. Prevention: Drowsiness-detection system warns drivers to prevent them falling asleep momentarily [Online]. Daimler. Available from: http://www.daimler.com/dccom/0-5-1210218-1-1210332-1-0-0-1210228-0-0-135-0-0-0-0-0-0-0.html. [Accessed 06.08.2012].

Daimon, T., Kawashima, H., 1996. New viewpoints for the evaluation of in-vehicle information systems: Applying methods in cognitive engineering. *JSAE Review*. 17, 151–157.

Damiani, S., Deregibus, E., Andreone, L., 2009. Driver-vehicle interfaces and interaction: Where are they going? *European Transport Research Review*. 1, 87–96.

Darwin, C.J., Turvey, M.T., Crowder, R.G., 1972. An auditory analogue of Sperling partial report procedure: Evidence for brief auditory storage. *Cognitive Psychology*. 3, 255–267.

References

Dehnig, W., Essig, H., Maass, S., 1981. *The adaptation of virtual man-computer interfaces to user requirements in dialogs*. Springer-Verlag, Berlin.

Department for Transport, 2011a. *Reported road casualties in Great Britain: Annual report 2010*. DfT, London.

Department for Transport, 2011b. *Strategic framework for road safety*. DfT, London.

Department for Transport, 2012. *Reported road casualties in Great Britain: Main results 2011*. DfT, London.

Dewar, R.E., Fenno, D., Garvey, P.M., Kuhn, B.T., Roberts, A.W., Schieber, F., Vincent, A., Yang, C.Y.D., Yim, Y.B., 2000. *User information systems: Developments and issues for the 21st century*. Transportation Research Board, Washington, DC.

de Winter, J.C.F., de Groot, S., Mulder, M., Wieringa, P.A., Dankelman, J., Mulder, J.A., 2009. Relationships between driving simulator performance and driving test results. *Ergonomics*. 52, 137–153.

Dillon, A., 2009. Inventing HCI: The grandfather of the field. *Interacting with Computers*. 21, 367–369.

Dingus, T.A., Klauer, S.G., Neale, V.L., Petersen, A., Lee, S.E., Sudweeks, J., Perez, M.A., Hankey, J., Ramsey, D., Gupta, S., Bucher, C., Doerzaph, Z.R., Jermeland, J., Knipling, R.R., 2006. *The 100-car naturalistic driving study, phase II—results of the 100-car field experiment*. National Highway Traffic Safety Administration, Washington, DC.

Dix, A., 2010. Human-computer interaction: A stable discipline, a nascent science, and the growth of the long tail. *Interacting with Computers*. 22, 13–27.

Doubleday, A., Ryan, M., Springett, M., Sutcliffe, A., 1997. A comparison of usability techniques for evaluating design. *Designing Interactive Systems Conference*, Amsterdam, August 18–20, 1997, ACM.

Douglas, S.A., Mithal, A.K., 1997. *The ergonomics of computer pointing devices*. Springer-Verlag, London.

Drews, F.A., Pasupathi, M., Strayer, D.L., 2008. Passenger and cell phone conversations in simulated driving. *Journal of Experimental Psychology: Applied*. 14, 392–400.

Driver, J., Bayliss, G.C., 1989. Movements and visual attention: The spotlight metaphor breaks down. Journal of Experimental Psychology: *Human Perception and Performance*. 15, 448–456.

Duchowski, A.T., 2007. *Eye tracking methodology theory and practice*. Springer-Verlag, London.

Dukic, T., Hanson, L., Holmqvist, K., Wartenberg, C., 2005. Effect of button location on driver's visual behaviour and safety perception. *Ergonomics*. 48, 399–410.

Dul, J., Bruder, R., Buckle, P., Carayon, P., Falzon, P., Marras, W.S., Wilson, J.R., van der Doelen, B., 2012. A strategy for human factors/ergonomics: Developing the discipline and profession. *Ergonomics*. 55, 377–395.

Dul, J., Weerdmeester, B., 2001. *Ergonomics for beginners: A quick reference guide*. Taylor & Francis, London.

Dzida, W., Herda, S., Itzfeldt, W.D., 1978. User-perceived quality of interactive systems. *3rd International Conference on Software Engineering*, Atlanta, GA, May 10–12, 2001, IEEE.

Eason, K.D., 1984. Towards the experimental study of usability. *Behaviour and Information Technology*. 3, 133–143.

Empeg, 2012. Empeg: Does your car stereo run Linux? [Online]. Available from: http://www.empeg.com/. [Accessed 11.05.2012].

Endsley, M.R., 1995. Toward a theory of situation awareness in dynamic systems. *Human Factors*. 37, 32–64.

Engström, J., Arfwidsson, J., Amditis, A., Andreone, L., Bengler, K., Cacciabue, P.C., Eschler, J., Nathan, F., Janssen, W., 2004. Meeting the challenges of future automotive HMI design: An overview of the AIDE integrated project. *ITS Congress*, Budapest, May 24, 2004, IEEE.

The European Conference of Ministers of Transport, 2003. *Statement of principles of good practice concerning the ergonomics and safety of in-vehicle information systems.* ECMT, Leipzig.

Fang, X., Xu, S., Brzezinski, J., Chan, S.S., 2006. A study of the feasibility and effectiveness of dual-mode information presentations. *International Journal of Human-Computer Interaction.* 20, 3–17.

Farago, R., 2002. BMW iDrive. *The Truth About Cars*, [Online]. The truth about cars. Available from: http://www.thetruthaboutcars.com/2002/02/bmw-i-drive/. [Accessed 21.06.2012].

Fastrez, P., Haué, J.-B., 2008. Editorial: Designing and evaluating driver support systems with the user in mind. *International Journal of Human-Computer Studies.* 66, 125–131.

Faurote, F.L., 1907. *A busy man's textbook on automobiles.* R R Donnelley and Sons, Chicago, IL.

Fitts, P.M., 1954. The information capacity of the human motor system in controlling the amplitude of movement. *Journal of Experimental Psychology: General.* 47, 381–391.

Fitts, P.M., Posner, M.I., 1967. *Human performance.* Brooks Cole, CA.

Flannagan, M., Sivak, M., 1993. Indirect vision systems, in: B. Peacock and W. Karwowski (eds.). *Automotive ergonomics.* Taylor & Francis, London, pp. 205–218.

Fleetwood, M.D., Byrne, M.D., 2006. Modeling the visual search of displays: A revised ACT-R model of icon search based on eye-tracking data. *Human-Computer Interaction.* 21, 153–197.

Fleischmann, T., 2007. Model based HMI specification in an automotive context, in: M.J. Smith and G. Salvendy (eds.). *Human interface and the management of information: Methods, techniques and tools in information design.* Springer-Verlag, Berlin, pp. 31–39.

Flemming, S.A.C., Hilliard, A., Jamieson, G.A., 2008. The need for human factors in the sustainability domain. *Proceedings of the Human Factors and Ergonomics Society Annual Meeting.* 52, 748–752.

Franks, I.M., Goodman, D., Miller, G., 1983. Human factors in sports systems: An empirical investigation of events in team games. *Proceedings of the Human Factors and Ergonomics Society Annual Meeting.* 27, 383–386.

Fuller, R., 2005. Towards a general theory of driver behaviour. *Accident Analysis and Prevention.* 37, 461–472.

Gelau, C., Schindhelm, R., 2010. Enhancing the occlusion technique as an assessment tool for driver visual distraction. *IET Intelligent Transport Systems.* 4, 346–355.

Goldstein, E.B., 2002. *Sensation and perception.* Wadsworth, Pacific Grove, CA.

Gould, J.D., Lewis, C., 1985. Designing for usability: Key principles and what designers think, in: P.J. Denning (ed.). *Communications of the ACM.* ACM, New York, pp. 300–311.

Graham-Rowe, D., 2010. *Touch screens that touch back.* Massachusetts Institute of Technology, Cambridge, MA.

Gray, W.D., 2002. Simulated task environments: The role of high-fidelity simulations, scaled worlds, synthetic environments, and laboratory tasks in basis and applied cognitive research. *Cognitive Science Quarterly.* 2, 205–227.

Gray, W.D., John, B.E., Atwood, M.E., 1993. Project Ernestine: Validating a GOMS analysis for predicting and explaining real-world task performance. *Human-Computer Interaction.* 8, 237–309.

Gray, W.D., Salzman, M.C., 1998. Damaged merchandise? A review of experiments that compare usability evaluation methods. *Human-Computer Interaction.* 13, 203–261.

Green, P., 1999. The 15-second rule for driver information systems. *ITS America Ninth Annual Meeting*, Washington, DC, April 9th, 1999, Intelligent Transportation Society of America.

References

Green, P., 2005. How driving simulator quality can be improved. *3rd Driving Simulation Conference North America*, Orlando, FL, November 2005, Elsevier.

Green, P., Hoekstra, E., Williams, M., 1993. *Further on-the-road tests of driver interfaces: Examination of a route guidance system and a car phone.* University of Michigan Transportation Research Institue (UMTRI), Ann Arbor, MI.

Green, P., Levison, W., Paelke, G., Serafin, C., 1994. *Suggested human factors design guidelines for driver information systems.* The University of Michigan Transportation Research Institute (UMTRI), Ann Arbor, MI.

Green, P., Levison, W., Paelke, G., Serafin, C., 1995. *Preliminary human factors design guidelines for driver information systems.* National Technical Information Service, Springfield, Virginia, VA.

Green, P., Tsimhoni, O., 2001. Visual occlusion to assess the demands of driving and tasks: *The literature. Workshop on Exploring the Occlusion Technique: Progress in Recent Research and Applications*, Fiat Research Centre, Torino, Italy, November 12–13, 2001, UMTRI.

Greenberg, S., Buxton, B., 2008. Usability evaluation considered harmful (some of the time). *CHI 2008 Conference on Human Factors in Computing Systems*, Florence, April 5–10, 2008, ACM.

Griffin, T.G.C., Young, M.S., Stanton, N.A., 2010. Investigating accident causation through information network modelling. *Ergonomics*. 53, 198–210.

Grudin, J., 2009. Brian Shackel's contribution to the written history of Human-Computer Interaction. *Interacting with Computers*. 21, 370–374.

Gu Ji, Y., Jin, B.S., 2010. Development of the conceptual prototype for haptic interface on the telematics system. *International Journal of Human-Computer Interaction*. 16, 22–52.

Hancock, P.A., Drury, C.G., 2011. Does human factors/ergonomics contribute to the quality of life? *Theoretical Issues in Ergonomics Science*. 12, 416–426.

Hancock, P.A., Hancock, G.M., Warm, J.S., 2009a. Individuation: The N = 1 revolution. *Theoretical Issues in Ergonomics Science*. 10, 481–488.

Hancock, P.A., Mouloua, M., Senders, J.W., 2009b. On the philosophical foundations of the distracted driver and driving distraction, in: M.A. Regan, J.D. Lee and K.L. Young (eds.). *Driver distraction: Theory, effects and mitigation*. CRC Press, Boca Raton, FL, pp. 11–30.

Hancock, P.A., Szalma, J.L., 2004. On the relevance of qualitative methods for ergonomics. *Theoretical Issues in Ergonomics Science*. 5, 499–506.

Harbluk, J.L., Noy, Y.I., Trbovich, P.L., Eizenman, M., 2007. An on-road assessment of cognitive distraction: Impacts on drivers' visual behavior and braking performance. *Accident Analysis and Prevention*. 39, 372–379.

Harrison, A., 1997. A survival guide to critical path analysis. Butterworth-Heinemann, Oxford, UK.

Hartson, H.R., 1998. HCI: Interdisciplinary roots and trends. *The Journal of Systems and Software*. 43, 103–118.

Harvey, C., Stanton, N.A., In press. The trade-off between context and objectivity in an analytic approach to the evaluation of in-vehicle interfaces. *IET Intelligent Transport Systems*. 6(3): 243–258.

Harvey, C., Stanton, N.A., Pickering, C.A., McDonald, M., Zheng, P., 2011a. Context of use as a factor in determining the usability of in-vehicle devices. *Theoretical Issues in Ergonomics Science*. 12, 318–338.

Harvey, C., Stanton, N.A., Pickering, C.A., McDonald, M., Zheng, P., 2011b. In-vehicle information systems to meet the needs of drivers. *International Journal of Human-Computer Interaction*. 27, 505–522.

Harvey, C., Stanton, N.A., Pickering, C.A., McDonald, M., Zheng, P., 2011c. To twist or poke? A method for identifying usability issues with the rotary controller and touch screen for control of in-vehicle information systems. *Ergonomics*. 54, 609–625.

Harvey, C., Stanton, N.A., Pickering, C.A., McDonald, M., Zheng, P., 2011d. A usability evaluation toolkit for in-vehicle information systems (IVISs). *Applied Ergonomics*. 42, 563–574.

Haslegrave, C.M., 1993. Visual aspects in vehicle design, in: B. Peacock and W. Karwowski (eds.). *Automotive ergonomics*. Taylor & Francis, London, pp. 79–98.

Hawthorn, D., 2000. Possible implications of aging for interface designers. *Interacting with Computers*. 12, 507–528.

Heaton, N.O., 1992. Defining usability. *Displays*. 13, 147–150.

Hedlund, J., Simpson, H., Mayhew, D., 2006. *International conference on distracted driving: Summary of proceedings and recommendations*. The Traffic Injury Research Foundation and The Canadian Automobile Association, Toronto.

Heide, A., Henning, K., 2006. The 'cognitive car': A roadmap for research issues in the automotive sector. *Annual Reviews in Control*. 30, 197–203.

Herriotts, P., 2005. Identification of vehicle design requirements for older drivers. *Applied Ergonomics*. 36, 255–262.

Herslund, M.-B., Jørgensen, N.O., 2003. Looked-but-failed-to-see-errors in traffic. *Accident Analysis and Prevention*. 35, 885–891.

Herzberg, F., 1996. *Work and the nature of man*. Staples Press, London.

Hewett, T.A., 1986. The role of iterative evaluation in designing systems for usability, in: M.D. Harrison and A.F. Monk (eds.). People and Computers: Designing for Usability. *Proceedings of the Second Conference of the British Computer Society Human Computer Special Interest Group*, University of York, September 23–26, 1986.

Hick, W.E., 1952. On the rate of gain of information. *Quarterly Journal of Experimental Psychology*. 4, 11–26.

Hix, D., Hartson, H.R., 1993. *Developing user interfaces: Ensuring usability through product and process*. John Wiley & Sons, Chichester.

Ho, C., Spence, C., 2008. *The multisensory driver: Implications for ergonomic car interface design*. Ashgate, Aldershot.

Hodgkinson, G.P., Crawshaw, C.M., 1985. Hierarchical task analysis for ergonomics research. *Applied Ergonomics*. 16, 289–299.

Hoedemaeker, M., Neerincx, M., 2007. Attuning in-car user interfaces to the momentary cognitive load, in: D.D. Schmorrow and L.M. Reeves (eds.). *Foundations of augmented cognition*. Springer-Verlag, Berlin, pp. 286–293.

Hoffman, J.E., 1998. Visual attention and eye movements, in: H. Pashler (ed.). *Attention*. Psychology Press, Hove, pp. 119–153.

Honda Motor Company, 2012. Honda history: A dynamic past, an exciting future [Online]. Honda. Available from: http://world.honda.com/history/. [Accessed 11.05.2012].

Horberry, T., Anderson, J., Regan, M.A., Triggs, T.J., Brown, J., 2006. Driver distraction: The effects of concurrent in-vehicle tasks, road environment complexity and age on driving performance. *Accident Analysis and Prevention*. 38, 185–191.

Hornbæk, K., 2006. Current practice in measuring usability. *International Journal of Human-Computer Studies*. 64, 79–102.

Horrey, W.J., 2011. Assessing the effects of in-vehicle tasks on driving performance. *Ergonomics in Design: The Quarterly of Human Factors Applications*. 19, 4–7.

Horrey, W.J., Alexander, A.L., Wickens, C.D., 2003. *Does workload modulate the effects of in-vehicle display location on concurrent driving and side task performance?* Aviation Human Factors Division, Institute of Aviation, University of Illinois, Savoy, IL.

Horrey, W.J., Wickens, C.D., 2007. In-vehicle glance duration: Distributions, tails and a model of crash risk. *Transportation Research Board*. 2018, 22–28.

Horrey, W.J., Wickens, C.D., Consalus, K.P., 2006. Modeling drivers' visual attention allocation while interacting with in-vehicle technologies. *Journal of Experimental Psychology: Applied.* 12, 67–78.
Howarth, P.A., 1991. Assessment of the visual environment, in: J.R. Wilson and E.N. Corlett (eds.). *Evaluation of human work: A practical ergonomics methodology.* Taylor & Francis, London, pp. 351–386.
Hulse, M.C., Dingus, T.A., Mollenhauer, M.A., Liu, Y., Jahns, S.K., Brown, T., McKinney, B., 1998. *Development of human factors guidelines for advanced traveler information systems and commercial vehicle operations: Identification of the strengths and weaknesses of alternative information display formats.* Federal Highway Administration, Washington, DC.
Hutchins, S.G., 2011. Evaluating a macrocognition model of team collaboration using real-world data from the Haiti relief effort. *Proceedings of the Human Factors and Ergonomics Society Annual Meeting.* 55, 252–256.
International Organization for Standardization, 1996. DD 235. Guide to in-vehicle information systems.
International Organization for Standardization, 1998. ISO 9241. Ergonomic requirements for office work with visual display terminals (VDTs)—part 11: Guidance on usability.
International Organization for Standardization, 1999. ISO 13407. Human-centred design processes for interactive systems.
International Organization for Standardization, 2001. ISO/IEC 9126. Information technology—software product quality—part 1: Quality model.
International Organization for Standardization, 2002. ISO 15007. Road vehicles—measurement of driver visual behaviour with respect to transport information and control systems—part 1: Definitions and parameters.
International Organization for Standardization, 2003. ISO 17287. Road vehicles—ergonomic aspects of transport information and control systems—procedure for assessing suitability of use while driving.
International Organization for Standardization, 2006. ISO 202082-1. Ease of operation of everyday products—part 1: Design requirements for context of use and user characteristics.
International Organization for Standardization, 2007. BS ISO 16673:2007. Road vehicles—ergonomic aspects of transport information and control systems—occlusion method to assess visual demand due to the use of in-vehicle systems.
Jaguar Cars Limited, 2011. Jaguar: Dual-view touch-screen [Online]. Available from: http://www.jaguarxjmedia.com/ENG_ENG_dualscreen.html#/dual-screen. [Accessed 03.08.2012].
Jamson, H., Merat, N., Carsten, O., Lai, F., 2011. Fully-automated driving: The road to future vehicle. *The 6th International Driving Symposium on Human Factors in Driver Assessment, Training and Vehicle Design*, Lake Tahoe, CA, June 27–30, 2011, TRB.
Jamson, H.A., Merat, N., 2005. Surrogate in-vehicle information systems and driver behaviour: Effects of visual and cognitive load in simulated rural driving. *Transportation Research Part F: Traffic Psychology and Behaviour.* 8, 79–96.
Japan Automobile Manufacturers Association, 2004. Guideline for in-vehicle display systems, version 3.0. JAMA, Tokyo.
Jastrzembski, T.S., Charness, N., 2007. The model human processor and the older adult: Parameter estimation and validation within a mobile phone task. *Journal of Experimental Psychology: Applied.* 13, 224–248.
Jeffries, R., Miller, J.R., Wharton, C., Uyeda, K.M., 1991. User interface evaluation in the real world: A comparison of four techniques. *CHI 1991 Conference on Human Factors in Computing Systems*, Pittsburgh, PA, May 15–20, 1999, ACM.

Johansson, E., Engström, J., Cherri, C., Nodari, E., Toffetti, A., Schindhelm, R., Gelau, C., 2004. Review of existing techniques and metrics for IVIS and ADAS assessment. Volvo Technology Corporation (VTEC), Gothenburg.

John, B.E., 1990. Extensions of GOMS analyses to expert performance requiring perception of dynamic visual and auditory information. *Conference on Human Factors in Computing Systems—CHI 1990*, Seattle, Washington, April 1–5, 1990, ACM.

John, B.E., 2011. Using predictive human performance models to inspire and support UI design recommendations. *CHI 2011 Conference on Human Factors in Computing Systems*, Vancouver, May 7–12, 2011, ACM.

John, B.E., Gray, W.D., 1995. CPM-GOMS: An analysis method for tasks with parallel activities. *Conference on Human Factors in Computing Systems—CHI 1995*, New York, May 7–11, 1995, ACM.

John, B.E., Jastrzembski, T.S., 2010. Exploration of costs and benefits of predictive human performance modeling for design. *International Conference on Cognitive Modeling*, Philadelphia, PA, August 6–8, 2004, Psychology Press.

John, B.E., Kieras, D.E., 1996. The GOMS family of user interface analysis techniques: Comparison and contrast. *ACM Transactions on Computer-Human Interaction*. 3, 320–351.

John, B.E., Newell, A., 1987. Predicting the time to recall computer command abbreviations. *Conference on Human Factors in Computing Systems and Graphics Interface—CHI/GI 1987*, Toronto, Canada, April 5–9, 1987, ACM.

John, B.E., Prevas, K., Salvucci, D.D., Koedinger, K., 2004a. Predictive human performance modeling made easy. *Conference on Human Factors in Computing Systems—CHI 2004*, Vienna, Austria, April 24–29, 2004, ACM.

John, B.E., Salvucci, D.D., Centgraf, P., Prevas, K., 2004b. Integrating models and tools in the context of driving and in-vehicle devices. *Sixth International Conference on Cognitive Modelling*, Pittsburgh, PA, July 30–August 1, 2004, Psychology Press.

Johnson, G.I., Clegg, C.W., Ravden, S.J., 1989. Towards a practical method of user interface evaluation. *Applied Ergonomics*. 20, 255–260.

Jokela, T., Iivari, N., Matero, J., Karukka, M., 2003. The standard of user-centered design and the standard definition of usability: Analyzing ISO 13407 against ISO 9241-11, in: C.S. De Souza, A. Sànchez, S. Barbosa and C. Gonzalez (eds.). *Proceedings of the 1st Latin American Conference on Human-Computer Interaction*, Rio de Janeiro, August 17–20, 2003.

Jordan, P.W., 1998a. Human factors for pleasure in product use. *Applied Ergonomics*. 29, 25–33.

Jordan, P.W., 1998b. *An introduction to usability*. Taylor & Francis, London.

Kadar, E.E., Effken, J.A., 2005. From discrete actors to goal directed actions: Toward a process-based methodology for psychology. *Philosophical Psychology*. 18, 353–382.

Kantowitz, B., 2000. In-vehicle information systems: Premises, promises, and pitfalls. *Transportation Human Factors*. 2, 359–379.

Kantowitz, B.H., 1992. Selecting measures for human factors research. *Human Factors*. 34, 387–398.

Karwowski, W., 2006. *International encyclopedia of ergonomics and human factors*. CRC Press, Boca Raton, FL.

Kass, S.J., Cole, K.S., Stanny, C.J., 2007. Effects of distraction and experience on situation awareness and simulated driving. *Transportation Research Part F: Traffic Psychology and Behaviour*. 10, 321–329.

Kern, D., Schmidt, A., 2009. Design space for driver-based automotive user interfaces. *First International Conference on Automotive User Interfaces and Interactive Vehicular Applications—Automotive UI*, Essen, Germany, September 21–22, 2009, ACM.

Khalid, H.M., Helander, M.G., 2004. A framework for affective customer needs in product design. *Theoretical Issues in Ergonomics Science*. 5, 27–42.

References

Khan, A.M., Bacchus, A., Erwin, S., 2012. Policy challenges of increasing automation in driving. *IATSS Research*. 35, 79–89.

Kieras, D., 2001. *Using the keystroke-level model to estimate execution times*. University of Michigan, Ann Arbor, MI.

Kieras, D.E., Meyer, D.E., 1997. An overview of the EPIC architecture for cognition and performance with application to human-computer interaction. *Human-Computer Interaction*. 12, 391–438.

Kirwan, B., 1992a. Human error identification in human reliability assessment. Part 1: Overview of approaches. *Applied Ergonomics*. 23, 299–318.

Kirwan, B., 1992b. Human error identification in human reliability assessment. Part 2: Detailed comparison of techniques *Applied Ergonomics*. 23, 371–381.

Kirwan, B., Ainsworth, L.K., 1992. *A guide to task analysis*. Taylor & Francis, London.

Klauer, S.G., Dingus, T.A., Neale, V.L., Sudweeks, J., Ramsey, D.J., 2006. *The impact of driver inattention on near-crash/crash risk: An analysis using the 100-car naturalistic driving study data*. Virginia Tech Transportation Institute, Blacksburg, VA.

Koninklijke Philips Electronics, 2012. Key Inventions [Online]. Available from: http://www.philips.com/about/company/history/keyinventions/index.page. [Accessed 15.05.2012].

Kontogiannis, T., Embrey, D., 1997. A user-centred design approach for introducing computer-based process information systems. *Applied Ergonomics*. 28, 109–119.

Kramer, A.F., McCarley, J.S., 2003. Oculomotor behaviour as a reflection of attention and memory processes: Neural mechanisms and applications to human factors. *Theoretical Issues in Ergonomics. Science*. 4, 21–55.

Kujala, T., Saariluoma, P., 2001. Effects of menu structure and touch screen scrolling style on the variability of glance durations during in-vehicle visual search tasks. *Ergonomics*. 54, 716–732.

Kurosu, M., 2007. Concept of usability revisited. *CHI 2007 Conference on Human-Computer Interaction*, Berlin, July 22–27, 2007.

Kwahk, J., Han, S.H., 2002. A methodology for evaluating the usability of audiovisual consumer electronic products. *Applied Ergonomics*. 33, 419–431.

Landau, K., 2002. Usability criteria for intelligent driver assistance systems. *Theoretical Issues in Ergonomics Science*. 3, 330–345.

Landauer, T.K., 1997. Behavioral research methods in human-computer interaction, in: M. Helander, T.K. Landauer and P. Prabhu (eds.). *Handbook of human-computer interaction*. Elsevier Science, Oxford, pp. 203–227.

Langham, M., Hole, G., Edwards, J., O'Neil, C., 2002. An analysis of 'looked but failed to see' accidents involving parked police vehicles. *Ergonomics*. 45, 167–185.

Lansdale, M.W., Ormerod, T.C., 1994. *Understanding interfaces: A handbook of human-computer dialogue*. Academic Press, London.

Lansdown, T.C., 2000. Driver visual allocation and the introduction of intelligent transport systems. *Proceedings of the Institution of Mechanical Engineers, part D: Journal of Automobile Engineering*. 214, 645–652.

Lansdown, T.C., Brook-Carter, N., Kersloot, T., 2002. Primary task disruption from multiple in-vehicle systems. *Intelligent Transportation Systems Journal*. 7, 151–168.

Lansdown, T.C., Brook-Carter, N., Kersloot, T., 2004a. Distraction from multiple in-vehicle secondary tasks: Vehicle performance and mental workload implications. *Ergonomics*. 47, 91–104.

Lansdown, T.C., Burns, P.C., Parkes, A.M., 2004b. Perspectives on occlusion and requirements for validation. *Applied Ergonomics*. 35, 225–232.

Lee, J., 1986. Letter to the editor: Conceptual subtleties of the percentile. *International Journal of Epidemiology*. 15, 141–142.

Lee, J.D., Young, K.L., Regan, M.A., 2009. Defining driver distraction, in: M.A. Regan, J.D. Lee and K.L. Young (eds.). *Driver distraction: Theory, effects and mitigation*. CRC Press, Boca Raton, FL, pp. 31–40.

Lees, M.N., Lee, J.D., 2007. The influence of distraction and driving context on driver response to imperfect collision warning systems. *Ergonomics*. 50, 1264–1286.

Lévesque, V., Oram, L., MacLean, K., Cockburn, A., Marchuk, N.D., Johnson, D., Colgate, J.E., Peshkin, M.A., 2011. Frictional widgets: Enhancing touch interfaces with programmable friction. *CHI 2011 Conference on Human Factors in Computing Systems*, Vancouver, May 7–12, 2011, ACM.

Lexus, 2012. Advanced pre-collision system (ACPS) with driver attention monitor [Online]. Lexus. Available from: http://www.lexus.com/models/LSh/features/safety/advanced_precollision_system_apcs_with_driver_attention_monitor.html#disclaimer_1. [Accessed 06.08.2012].

Lindgaard, G., 2009. Early traces of usability as a science and as a profession. *Interacting with Computers*. 21, 350–352.

Lindgaard, G., Whitfield, T.W.A., 2004. Integrating aesthetics within an evolutionary and psychological framework. *Theoretical Issues in Ergonomics Science*. 5, 73–90.

Liu, C.-C., Doong, J.-L., Hsu, C.-C., Huang, W.-S., Jeng, M.-C., 2009. Evidence for the selective attention mechanism and dual-task interference. *Applied Ergonomics*. 40, 341–347.

Liu, Y.-C., 2000. Effect of advanced traveler information displays on younger and older drivers' performance. *Displays*. 21, 161–168.

Liu, Y.-C., Bligh, T., Chakrabarti, A., 2003. Towards an 'ideal' approach for concept generation. *Design Studies*. 24, 341–355.

Llaneras, R.E., Singer, J.P., 2002. *Inventory of in-vehicle technology human factors design characteristics*. National Highway Traffic Safety Administration, Washington, DC.

Lockhart, T.E., Shi, W., 2010. Effects of age on dynamic accommodation. *Ergonomics*. 53, 892–903.

Lockyer, K., 1984. *Critical path analysis and other project network techniques*. Pitman, London, UK.

Long, J., 2010. Some celebratory HCI reflections on a celebratory HCI festschrift. *Interacting with Computers*. 22, 68–71.

Long, J.B., 1986. People and computers: Designing for usability, in: M.D. Harrison and A.F. Monk (eds.). *2nd Conference of the British Computer Society Human Computer Interaction Specialist Group*, University of York, September 23–26, 1986.

Ludvigsen, K.E., 1997. A century of automobile comfort and convenience, in: Society of Automotive Engineers Historical Committee. Committee (ed.). *The automobile: A century of progress. Society of Automotive Engineers (SAE)*, Pennsylvania, pp. 99–120.

Lyons, M., 2009. Towards a framework to select techniques for error prediction: Supporting novice users in the healthcare sector. *Applied Ergonomics*. 40, 379–395.

Ma, R., Kaber, D.B., 2007. Situation awareness and driving performance in a simulated navigation task. *Ergonomics*. 50, 1351–1364.

Macdonald, A.S., 1998. Developing a quantitative sense, in: N. Stanton (ed.). *Human factors in consumer products*. Taylor & Francis, London, pp. 175–191.

Mackay, W.E., 2004. The interactive thread: Exploring methods for multi-disciplinary design. *5th Conference on Designing Interactive Systems: Processes, Practices, Methods, and Techniques*, Cambridge, MA, August 1–4, 2004, ACM.

Maguire, M., Kirakowski, J., Vereker, N., 1998. *RESPECT: User centred requirements handbook*. Ergonomics and Safety Research Institute (ESRI), Loughborough.

Manes, D., Green, P., Hunter, D., 1997. *Prediction of destination entry and retrieval times using keystroke-level models*. The University of Michigan Transportation Research Institute (UMTRI), Ann Arbor, MI.

Marcus, A., January–February 2004. Vehicle user interfaces: The next revolution. *Interactions*, 40–47.
Matthews, M.L., Bryant, D.J., Webb, R.D.G., Harbluk, J.L., 2001. Model for situation awareness and driving: Application to analysis and research for intelligent transportation systems. *Transportation Research Record*. 1779, 26–32.
McClelland, I., 1991. Product assessment and user trials, in: J.R. Wilson and E.N. Corlett (eds.). *Evaluation of human work: A practical ergonomics methodology*. Taylor & Francis, London, pp. 218–247.
McMains, S.A., Somers, D.C., 2004. Multiple spotlights of attentional selection in human visual cortex. *Neuron*. 42, 677–686.
Meister, D., 1989. *Conceptual aspects of human factors*. John Hopkins University Press, Baltimore, MD.
Meister, D., 1992. Some comments on the future of ergonomics. *International Journal of Industrial Ergonomics*. 10, 257–260.
Mendoza, P.A., Angelelli, A., Lindgren, A., 2011. Ecological interface design inspired human machine interface for advanced driver assistance systems. *IET Intelligent Transport Systems*. 5, 53–39.
Monk, C.A., Kidd, D.G., 2007. R we fooling ourselves: Does the occlusion technique shortchange R estimates? *Fourth International Driving Symposium on Human Factors in Driver Assessment, Training and Vehicle Design*, Stevenson, WA, July 9–12, 2007, TRB.
Moran, J., 2009. Paradise lost [Online]. BBC Magazine. Available from: http://news.bbc.co.uk/1/hi/magazine/8133890.stm. [Accessed Tuesday 7th July, 2009].
Motorola Solutions, 2012. A legacy of innovation: Timeline of Motorola history since 1928 [Online]. Available from: http://www.motorola.com/staticfiles/Business/Corporate/US-EN/history/feature-car-radio.html. [Accessed 11.05.2012].
Mourant, R.R., Herman, M., Moussa-Hamouda, E., 1980. Direct looks and control location in automobiles. *Human Factors*. 22, 417–425.
Mourant, R.R., Rockwell, T.H., 1970. Mapping eye-movement patterns to the visual scene in driving: An exploratory study. *Human Factors*. 12, 81–87.
Muller, M.M., Malinowski, P., Gruber, T., Hillyard, S.A., 2003. Sustained division of the attentional spotlight. *Nature*. 424, 309–312.
Munipov, V.M., 1992. Chernobyl operators: Criminals or victims? *Applied Ergonomics*. 23, 337–342.
Murdock, B.B., 1961. Short-term retention of single paired-associates. *Journal of Experimental Psychology*. 65, 433–443.
National Highway Traffic Safety Administration, 2012. What is distracted driving? [Online]. NHTSA. Available from: http://www.distraction.gov/content/get-the-facts/facts-and-statistics.html. [Accessed 02.08.2012].
Newcomb, T.P., Spurr, R.T., 1989. *A technical history of the motor car*. Adam Hilger, Bristol.
Nielsen, J., 1992. Finding usability problems through heuristic evaluation. *CHI 1992 Conference on Human Factors in Computing Systems*, Monterey, CA, May 3–7, 1992, ACM.
Nielsen, J., 1993. *Usability engineering*. Academic Press, London.
Nielsen, J., 2005. Ten usability heuristics [Online]. Available from: http://www.useit.com/papers/heuristic/heuristic_list.html. [Accessed 20.03.09].
Nielsen, J., Nov–Dec 2006. Logic versus usage: The case for activity-centered design. *Interactions*, 45.
Nielsen, J., 2009. Usability 101: Introduction to usability [Online]. Available from: http://www.useit.com/alertbox/20030825.html. [Accessed 22 May, 2009].
Nielsen, J., Phillips, V.L., 1993. Estimating the relative usability of two interfaces: Heuristic, formal, and empirical methods compared. *CHI 1993 Conference on Human Factors in Computing Systems*, Amsterdam, April 24–29, 1993, ACM.

Noel, E., Nonnecke, B., Trick, L., 2005. A comprehensive learnability evaluation method for in-car navigation devices. Society of Automotive Engineers, Washington, DC.

Nokia, 2009. History of Nokia part one: Nokia firsts [Online]. Available from: http://conversations.nokia.com/2009/01/19/history-of-nokia-part-one-nokia-firsts/. [Accessed 15.05.2012].

Norman, D.A., 1983. Design principles for human computer interfaces. *SIGCHI Conference on Human Factors in Computing Systems*, Boston, MA, December 12–15, 1983, ACM.

Norman, D.A., 2002. *The design of everyday things*. Basic Books, New York.

Norman, D.A., Draper, S.W., 1986. *User centered system design: New perspectives on human-computer interaction*. Lawrence Erlbaum, London.

Nowakowski, C., Green, P., 2001. Prediction of menu selection times parked and while driving using the SAE J2365 method. University of Michigan Transportation Research Institute (UMTRI), Ann Arbor, MI.

Nowakowski, C., Utsui, Y., Green, P., 2000. *Navigation system destination entry: The effects of driver workload and input devices, and implications for SAE recommended practice*. University of Michigan Transportation Research Institute (UMTRI), Ann Arbor, MI.

Noy, Y.I., Lemoine, T.L., Klachan, C., Burns, P.C., 2004. Task interruptability and duration as a measure of visual distraction. *Applied Ergonomics*. 35, 207–213.

Noyes, J., 2009. Telescreens, keypens, and the expert: A 60 year snapshot. *Interacting with Computers*. 21, 335–338.

Oliver, K.J., Burnett, G.E., 2008. Learning-oriented vehicle navigation systems: A preliminary investigation in a driving simulator. *Mobile HCI*, Amsterdam, September 2–5, 2008, ACM.

Olson, J.R., Olson, G.M., 1990. The growth of cognitive modeling in human-computer interaction since GOMS. *Human-Computer Interaction*. 5, 221–265.

Olson, J.S., Moran, T.P., 1996. Mapping the method muddle: Guidance in using methods for user interface design, in: M. Rudisill, C. Lewis, P.G. Polson and T.D. Mackay (eds.). *Human-computer interface designs: Success stories, emerging methods, and real world context*. Kaufmann, San Francisco, pp. 269–300.

Olson, P.L., 1993. Vision and perception, in: B. Peacock and W. Karwowski (eds.). *Automotive ergonomics*. Taylor & Francis, London, pp. 161–184.

Pashler, H.E., 1998. *The psychology of attention*. MIT Press, Cambridge, MA.

Pauzié, A., 2008. Evaluating driver mental workload using the driving activity load index (DALI). *1st European Conference on Human Centred Design for Intelligent Transport Systems*, Lyon, 3–4 April, 2008, INRETS.

Peacock, B., Karwowski, W., 1993. *Automotive ergonomics*. Taylor & Francis, London.

Peterson, L.R., Peterson, M.J., 1959. Short-term retention of individual verbal items. *Journal of Experimental Psychology*. 58, 193–198.

Pettitt, M., Burnett, G., Stevens, A., 2005. Defining driver distraction. *World Congress on Intelligent Transport Systems*, San Francisco, CA, November 6–10, 2005, Curran Associates.

Pettitt, M., Burnett, G., Stevens, A., 2007. An extended keystroke level model (KLM) for predicting the visual demand of in-vehicle information systems. *CHI 2007 Conference on Human Factors in Computing Systems*, San Jose, CA, April 28–May 3, 2007, ACM.

Pheasant, S., 1996. *Bodyspace: Anthropometry, ergonomics, and the design of work*. CRC Press, London, UK.

Pickering, C.A., Burnham, K.J., Richardson, M.J., 2007. A review of automotive human machine interface technologies and techniques to reduce driver distraction. *2nd IET International Conference on System Safety*, London, UK, October 22–24, 2007, IET.

Piechulla, W., Mayser, C., Gehrke, H., König, W., 2003. Reducing drivers' mental workload by means of an adaptive man-machine interface. *Transportation Research Part F: Traffic Psychology and Behaviour*. 6, 233–248.

References

Pioneer Europe, 2012. History [Online]. Available from: http://www.pioneer.eu/uk/content/company/company/history.html. [Accessed 14.05.2012].

Plant, K.L., Stanton, N.A., 2012. Why did the pilots shut down the wrong engine? Explaining errors in context using schema theory and the perceptual cycle model. *Safety Science.* 50, 300–315.

Posner, M.I., Petersen, S.E., 1990. The attention system of the human brain. *Annual Review of Neuroscience.* 13, 25–42.

Preece, J., Rogers, Y., Sharp, H., 2002. *Interaction design: Beyond human-computer interaction.* John Wiley & Sons, New Jersey.

Rakotonirainy, A., Tay, R., 2004. In-vehicle ambient intelligent transport systems (I-VAITS): Towards an integrated research. *7th International IEEE Conference on Intelligent Transportation Systems*, Washington, DC, October 3–6, 2004, IEEE.

Rantanen, E.M., Goldberg, J.H., 1999. The effect of mental workload on the visual field size and shape. *Ergonomics.* 42, 816–834.

Ravden, S., Johnson, G., 1989. *Evaluating usability of human-computer interfaces: A practical method.* Ellis Horwood, Chichester.

Reed, M.P., Green, P.A., 1999. Comparison of driving performance on-road and in a low-cost simulator using a concurrent telephone dialling task. *Ergonomics.* 42, 1015–1037.

Reed-Jones, J., Trick, L.M., Matthews, M., 2008. Testing assumptions implicit in the use of the 15-second rule as an early predictor of whether an in-vehicle device produces unacceptable levels of distraction. *Accident Analysis and Prevention.* 40, 628–634.

Regan, M.A., Lee, J.D., Young, K.L., 2009. Introduction, in: M.A. Regan, J.D. Lee and K.L. Young (eds.). *Driver distraction: Theory, effects and mitigation.* CRC Press, FL, pp. 3–7.

Reimer, B., Mehler, B., Coughlin, J.F., Roy, N., Dusek, J.A., 2011. The impact of a naturalistic hands-free cellular phone task on heart rate and simulated driving performance in two age groups. *Transportation Research part F: Traffic Psychology and Behaviour.* 14, 13–25.

Rennie, A.M., 1981. The application of ergonomics to consumer product evaluation. *Applied Ergonomics.* 12, 163–168.

Richardson, J.F., Pincherle, G., 1969. Heights and weights of British businessmen. *British Journal of Preventative and Social Medicine.* 23, 267–270.

Richter, H., Ecker, R., Deisler, C., Butz, A., 2010. HapTouch and the 2+1 state model: Potentials of haptic feedback on touch based in-vehicle information systems. *Automotive UI 2nd International Conference on Automotive User Interfaces and Interactive Vehicular Applications*, Pittsburgh, PA, November 11–12, 2010, University of Pittsburgh.

Rogé, J., Pébayle, T., Kiehn, L., Muzet, A., 2002. Alteration of the useful visual field as a function of state of vigilance in simulated car driving. *Transportation Research Part F: Traffic Psychology and Behaviour.* 5, 189–200.

Rogé, J., Pébayle, T., Lambilliotte, E., Spitzenstetter, F., 2004. Influence of age, speed and duration of monotonous driving task in traffic on the driver's useful field of view. *Vision Research.* 44, 2737–2744.

Rogers, W.A., Fisk, A.D., McLaughlin, A.C., Pak, R., 2005. Touch a screen or turn a knob: Choosing the best device for the job. *Human Factors.* 47, 271–288.

Rosenbaum, S., Skinner, R.K., 1985. A survey of heights and weights of adults in Great Britain, 1980. *Annals of Human Biology.* 12, 115–127.

Rudin-Brown, C.M., 2010. 'Intelligent' in-vehicle intelligent transport systems: Limiting behavioural adaptation through adaptive design. *IET Intelligent Transport Systems.* 4, 252–261.

Rumar, K., 1993. Road user needs, in: A.M. Parkes and S. Franzén (eds.). *Driving future vehicles.* Taylor & Francis, London, pp. 41–48.

Ryan, G.A., Legge, M., Rosman, D., 1998. Age-related changes in drivers' crash risk and crash type. *Accident Analysis and Prevention.* 30, 379–387.

Rydström, A., Bengtsson, P., Grane, C., Broström, R., Agardh, J., Nilsson, J., 2005. Multifunctional systems in vehicles: A usability evaluation. *CybErg 2005 International Cyberspace Conference on Ergonomics*, Johannesburg, September 15–October 15, 2005, International Ergonomics Association Press.

Salvucci, D.D., 1999. Mapping eye movements to cognitive processes. Doctoral thesis, Carnegie Mellon University, Pittsburgh, PA.

Salvucci, D.D., 2000. A model of eye movements and visual attention, in: N. Taatgen and J. Aasman (eds.). *Third International Conference on Cognitive Modeling*, Veenendaal, The Netherlands, March 23–25, 2000.

Salvucci, D.D., 2001. Predicting the effects of in-car interface use on driver performance: An integrated model approach. *International Journal of Human-Computer Studies*. 55, 85–107.

Salvucci, D.D., Zuber, M., Beregovaia, E., Markley, D., 2005. Distract-R: Rapid prototyping and evaluation of in-vehicle interfaces. *CHI 2005 Conference on Human Factors in Computing Systems*, Portland, OR, April 2–7, 2005.

Santos, J., Merat, N., Mouta, S., Brookhuis, K., deWaard, D., 2005. The interaction between driving and in-vehicle systems: Comparison of results from laboratory, simulator and real-world studies. *Transportation Research part F: Traffic Psychology and Behaviour*. 8, 135–146.

Sarter, N.B., 2006. Multimodal information presentation: Design guidance and research challenges. *International Journal of Industrial Ergonomics*. 36, 439–445.

Sauer, J., Seibel, K., Rüttinger, B., 2010. The influence of user expertise and prototype fidelity in usability tests. *Applied Ergonomics*. 41, 130–140.

Sauer, J., Sonderegger, A., 2009. The influence of prototype fidelity and aesthetics of design in usability tests: Effects on user behaviour, subjective evaluation and emotion. *Applied Ergonomics*. 40, 670–677.

Savelsbergh, G.J.P., Van der Kamp, J., Williams, A.M., Ward, P., 2005. Anticipation and visual search behaviour in expert soccer goalkeepers. *Ergonomics*. 48, 1686–1697.

Savoy, A., Guo, Y., Salvendy, G., 2009. Effects of importance and detectability of usability problems on sample size requirements. *International Journal of Human-Computer Interaction*. 25, 430–440.

Schiffman, H.R., 2001. *Sensation and perception: An integrated approach*. John Wiley & Sons, New York.

Schoelles, M.J., Gray, W.D., 2000. Argus prime: Modeling emergent microstrategies in a complex, simulated task environment. *3rd International Conference on Cognitive Modeling*, Veenendal, NL, Universal Press.

Senders, J.W., Kristofferson, A.B., Levison, W.H., Dietrich, C.W., Ward, J.L., 1967. The attentional demand of automobile driving. *Highway Research Record*. 1967, 15–33.

Seppelt, B., Wickens, C.D., 2003. *In-vehicle tasks: Effects of modality, driving relevance, and redundancy*. Aviation Human Factors Division, University of Illinois, Urbana-Champaign, IL.

Shackel, B., 1959. Ergonomics for a computer. *Design*. 120, 36–39.

Shackel, B., 1986. Ergonomics in design for usability. *2nd Conference of The British Computer Society Human Computer Special Interest Group*, York, September 23–26, 1986, University Press.

Shackel, B., 1997. Human-computer interaction—whence and whither? *Journal of the American Society for Information Science*. 48, 970–986.

Shinar, D., 2008. Looks are (almost) everything: Where drivers look to get information. *Human Factors*. 50, 380–384.

Shneiderman, B., 1992. *Designing the user interface: Strategies for effective human-computer interaction*. Addison-Wesley, New York.

References

Shneiderman, B., 2000. Universal usability: Pushing human-computer interaction research to empower every citizen. *Communications of the ACM.* 43, 84–91.

Sivak, M., 1996. The information that drivers use: Is it indeed 90% visual? *Perception.* 25, 1081–1089.

Snowden, R., Thompson, P., Troscianko, T., 2006. *Basic vision: An introduction to visual perception.* University Press, Oxford.

Snyder, R.G., Spencer, M.L., Owings, C.L., Schneider, L.W., 1975. *Physical characteristics of children: As related to death and injury for consumer product design and use.* University of Michigan Transportation Research Institute (UMTRI), Ann Arbor, MI.

Society of Automotive Engineers, 2002. SAE J2365 (proposed draft). Calculation of the time to complete in-vehicle navigation and route guidance tasks.

Sodhi, M., Reimer, B., Llamazares, I., 2002. Glance analysis of driver eye movements to evaluate distraction. *Behavior Research Methods Instruments and Computers.* 34, 529–538.

Sonderegger, A., Sauer, J., 2009. The influence of laboratory set-up in usability tests: Effects on user performance, subjective ratings and physiological measures. *Ergonomics.* 52, 1350–1361.

Sperling, G., 1960. The information available in brief visual presentations. *Psychological Monographs: General and Applied.* 74, 1–29.

Stanton, N.A., 1998. Product design with people in mind, in: N. Stanton (ed.). *Human factors in consumer products.* Taylor & Francis, London, pp. 1–17.

Stanton, N.A., 2006. Hierarchical task analysis: Developments, applications, and extensions. *Applied Ergonomics.* 37, 55–79.

Stanton, N.A., Baber, C., 1992. Usability and EC directive 90/270. *Displays.* 13, 151–160.

Stanton, N.A., Baber, C., 2008. Modelling of human alarm handling response times: A case study of the Ladbroke Grove rail accident in the UK. *Ergonomics.* 51, 423–440.

Stanton, N.A., Dunoyer, A., Leatherland, A., 2011. Detection of new in-path targets by drivers using stop and go adaptive cruise control. *Applied Ergonomics.* 42, 592–601.

Stanton, N.A., Salmon, P.M., 2009. Human error taxonomies applied to driving: A generic driver error taxonomy and its implications for intelligent transport systems. *Safety Science.* 47, 227–237.

Stanton, N.A., Salmon, P.M., Walker, G.H., Baber, C., Jenkins, D.P., 2005. *Human factors methods: A practical guide for engineering and design.* Ashgate, Aldershot.

Stanton, N.A., Stevenage, S.V., 1998. Learning to predict human error: Issues of acceptability, reliability and validity. *Ergonomics.* 41, 1737–1756.

Stanton, N.A., Walker, G.H., 2011. Exploring the psychological factors involved in the Ladbroke Grove rail accident. *Accident Analysis and Prevention.* 43, 1117–1127.

Stanton, N.A., Walker, G.H., Young, M.S., Kazi, T., Salmon, P.M., 2007a. Changing drivers' minds: The evaluation of an advanced driver coaching system. *Ergonomics.* 50, 1209–1234.

Stanton, N.A., Young, M., McCaulder, B., 1997. Drive-by-wire: The case of driver workload and reclaiming control with adaptive cruise control. *Safety Science.* 27, 149–159.

Stanton, N.A., Young, M.S., 1998a. Ergonomics methods in consumer product design and evaluation, in: N.A. Stanton (ed.). *Human factors in consumer products.* Taylor & Francis, London, pp. 21–54.

Stanton, N.A., Young, M.S., 1998b. Is utility in the mind of the beholder? A study of ergonomics methods. *Applied Ergonomics.* 29, 41–54.

Stanton, N.A., Young, M.S., 1999a. *A guide to methodology in ergonomics: Designing for human use.* Taylor & Francis, London.

Stanton, N.A., Young, M.S., 1999b. What price ergonomics? *Nature.* 399, 197–198.

Stanton, N.A., Young, M.S., 2003. Giving ergonomics away? The application of ergonomics methods by novices. *Applied Ergonomics.* 34, 479–490.

Stanton, N.A., Young, M.S., Walker, G.H., 2007b. The psychology of driving automation: A discussion with Professor Don Norman. *International Journal of Vehicle Design*. 45, 289–306.

Staunstrup, P., 2012. Mobile systems [Online]. Available from: http://www.ericssonhistory.com/templates/Ericsson/Article.aspx?id=2095&ArticleID=1390&CatID=363&epslanguage=EN. [Accessed 15.05.2012].

Sternberg, S., 1975. Memory scanning: New findings and current controversies. *Quarterly Journal of Experimental Psychology*. 27, 1–32.

Stevens, A., 2012. The future of in-vehicle navigation systems. *ITS International* [Online]. Route One Publishing Ltd. Available from: http://www.itsinternational.com/categories/location-based-systems/features/the-future-of-in-vehicle-navigation-systems/. [Accessed 01.08.2012].

Stevens, A., Board, A., Allen, P., Quimby, A., 1999. A safety checklist for the assessment of in-vehicle information systems: A user's manual. Transport Research Laboratory, London.

Stevens, A., Cynk, S., Beesley, R., 2011. Revision of the checklist for the assessment of in-vehicle information systems. Transport Research Laboratory, London.

Stevens, A., Quimby, A., Board, A., Kersloot, T., Burns, P., 2002. Design guidelines for safety of in-vehicle information systems. Transport Research Laboratory, London.

Strayer, D.L., Drews, F.A., 2007. Multitasking in the automobile, in: A.F. Kramer (ed.). *Attention: From theory to practice*. University Press, Oxford, pp. 121–133.

Summala, H., Nieminen, T., Punto, M., 1996. Maintaining lane position with peripheral vision during in-vehicle tasks. *Human Factors*. 38, 442–451.

Sutter, C., 2007. Sunsumotor transformation of input devices and the impact on practice and task difficulty. *Ergonomics*. 50, 1999–2016.

Sweeney, M., Maguire, M., Shackel, B., 1993. Evaluating user-computer interaction: A framework. *International Journal of Man-Machine Studies*. 38, 689–711.

Taveira, A.D., Choi, S.D., 2009. Review study of computer input devices and older users. *International Journal of Human-Computer Interaction*. 25, 455–474.

Tijerina, L., Parmer, E., Goodman, M.J., 1998. Driver workload assessment of route guidance system destination entry while driving: A test track study. *5th ITS World Congress*, Seoul, October12–16, 1998, ITS.

Tilley, A.R., Henry Dreyfuss Associates, 2002. *The measure of man and woman: Human factors in design*. John Wiley & Sons, New York.

Tingvall, C., Eckstein, L., Hammer, M., 2009. Government and industry perspectives on driver distraction, in: M.A. Regan, J.D. Lee and K.L. Young (eds.). *Driver distraction: Theory, effects and mitigation*. CRC Press, Boca Raton, FL, pp. 603–618.

Tsimhoni, O., Green, P., 2003. Time-sharing of a visual in-vehicle task while driving: The effects of four key constructs. *2nd International Driving Symposium on Human Factors in Driver Assessment, Training and Vehicle Design*, Park City, Utah, July 21–24, 2003, Elsevier.

Tsimonhi, O., Smith, D., Green, P., 2004. Address entry while driving: Speech recognition versus a touch-screen keyboard. *Human Factors*. 46, 600–610.

Tsimhoni, O., Yoo, H., Green, P., 1999. *Effects of visual demand and in-vehicle task complexity on driving and task performance as assessed by visual occlusion*. University of Michigan Transportation Research Instutute (UMTRI), Ann Arbor, MI.

van der Horst, R., 2004. Occlusion as a measure for visual workload: An overview of TNO occlusion research in car driving. *Applied Ergonomics*. 35, 189–196.

Victor, T.W., Engström, J., Harbluck, J.L., 2009. Distraction asessment methods based on visual behaviour and event detection, in: M.A. Regan, J.D. Lee and K.L. Young (eds.). *Driver distraction: Theory, effects, and mitigation*. CRC Press, Boca Raton, FL, pp. 135–165.

Volkswagen, 2012. Passat saloon: Driver-assistance [Online]. Volkswagen. Available from: http://www.volkswagen.co.uk/#/new/passat-vii/explore/experience/driver-assistance/. [Accessed 06.08.2012].

Walker, G.H., Stanton, N.A., Young, M.S., 2001. Where is computing driving cars? *International Journal of Human-Computer Interaction.* 13, 203–229.

Walter, S., 1986. Letter to the editor: Problems with percentiles. *International Journal of Epidemiology.* 15, 431–432.

Wang, X., Trasbot, J., 2011. Effects of target location, stature and hand grip type on in-vehicle reach discomfort. *Ergonomics.* 54, 466–476.

Wang, Y., Mehler, B., Reimer, B., Lammers, V., D'Ambrosio, L.A., Coughlin, J.F., 2010. The validity of driving simulation for assessing differences between in-vehicle informational interfaces: A comparison with field testing. *Ergonomics.* 53, 404–420.

Whitfield, D., Langford, J., 2001. *The Oxford companion to the body: No. 1.* Oxford University Press, Oxford.

Wickens, C.D., 1991. Processing resources and attention in: D.L. Damos (ed.). *Multiple-task performance.* Taylor & Francis, London, pp. 3–34.

Wickens, C.D., 2002. Multiple resources and performance prediction. *Theoretical Issues in Ergonomics Science.* 3, 159–177.

Wierwille, W.W., 1993. Visual and manual demands of in car controls and displays, in: B. Peacock and B. Karwowski (eds.). *Automotive ergonomics.* Taylor & Francis, London, UK, pp. 299–313.

Wilkinson, L., 1999. Statistical methods in psychology journals: Guidelines and explanations. *American Psychologist.* 54, 594–604.

Wilson, C.E., 2006. Triangulation: The explicit use of multiple methods, measures, and approaches for determining core issues in product development. *Interactions.* 46–63.

Wilson, J.R., Corlett, N., 2005. *Evaluation of human work.* CRC Press, Boca Raton, FL.

Wittmann, M., Kiss, M., Gugg, P., Steffen, A., Fink, M., Pöppel, E., Kamiya, H., 2006. Effects of display position of a visual in-vehicle task on simulated driving. *Applied Ergonomics.* 37, 187–199.

Wolfe, J.M., 1998. Visual search, in: H. Pashler (ed.). *Attention.* Psychology Press, Hove, pp. 13–73.

Wolfe, J.M., 2007. Guided search 4.0: Current progress with a model of visual search, in: W. Gray (ed.). *Intergrated models of cognitive systems.* Oxford University Press, New York, pp. 99–119.

World Health Organization, 2011. Mobile phone use: A growing problem of driver distraction. WHO, Geneva, Switzerland.

Young, K.L., Lenné, M.G., Williamson, A.R., 2011a. Sensitivity of the lane change test as a measure of in-vehicle system demand. *Applied Ergonomics.* 42, 611–618.

Young, K.L., Regan, M.A., Hammer, M., 2003. Driver distraction: A review of the literature. Monash University Accident Research Centre, Victoria.

Young, K.L., Regan, M.A., Lee, J.D., 2009a. Factors moderating the impact of distraction on driving performance and safety, in: M.A. Regan, J.D. Lee and K.L. Young (eds.). *Driver distraction: Theory, effects and mitigation.* CRC Press, Boca Raton, FL, pp. 335–351.

Young, K.L., Regan, M.A., Lee, J.D., 2009b. Measuring the effects of driver distraction: Direct driving performance methods and measures, in: M.A. Regan, J.D. Lee and K.L. Young (eds.). *Driver distraction: Theory, effects and mitigation.* CRC Press, Boca Raton, FL, pp. 85–105.

Young, M.S., Birrell, S.A., Stanton, N.A., 2011b. Safe driving in a green world: A review of driver performance benchmarks and technologies to support 'smart' driving. *Applied Ergonomics.* 42, 533–539.

Young, M.S., Mahfoud, J.M., Walker, G.H., Jenkins, D.P., Stanton, N.A., 2008. Crash dieting: The effects of eating and drinking on driving performance. *Accident Analysis and Prevention*. 40, 142–148.

Young, M.S., Stanton, N.A., 2002. Malleable attentional resources theory: A new explanation for the effects of mental underload on performance. *Human Factors*. 44, 365–375.

Young, M.S., Stanton, N.A., 2007. What's skill got to do with it? Vehicle automation and driver mental workload. *Ergonomics*. 50, 1324–1339.

Index

A

Analytic evaluation, 52, 71–101
    data analysis, 76
    equipment, 72–74
        data collection apparatus, 74
        IVIS, 72–74
    heuristic analysis, 59, 92–94
        for IVIS evaluation, 92
        utility, 92–94
    hierarchical task analysis, 57–58, 77–79
        for IVIS evaluation, 77–79
        utility, 79
    layout analysis, 59–60, 94–96
        for IVIS evaluation, 94–96
        utility, 96
    methods, 57–60, 72–76
    multimodal CPA, 58, 80–87
        CPA for IVIS evaluation, 83–87
        CPA utility, 87
        defining CPA activities, 80–83
    procedure, 74–76
    systematic human error reduction, prediction approach, 58–59, 87–92
        for IVIS evaluation, 87–91
        utility, 92
Attention, visual, 153–173
    CPA model limitations, 171–172
    design implications, 171
    dual-task IVIS interaction, CPA model, 156–158
        dual-task CPA calculator, 158
        glance behaviour data, 156
        median glance times, 158
        model assumptions, 157
        visual information processing, 157–158
        visual mode, 157
        visual operation initiation, 158
    glance behaviour, dual-task environment, 159–165
        chunking of visual information, 163
        sequential operations, 163–165
        shared glances, 163
        shared visual attention, 159–163
    glance behaviour data analysis, 155–156
    multidimensionality, 167–168
    occlusion technique, 169
    road, glance durations, 169–171
    shared glance CPA model, 165–167
    visual behaviour, driving, 154–155
        dual-task glance durations, 154–155
    visual behaviour theory, driving, 168–169
Automotive manufacturers, 178

C

Comfort control, 8
    trends in, 9
Computer-human interaction
    ergonomics, 12–14
    modelling, 133–136
        task times, 134
        techniques, 134–136
Context, objectivity, distinguished, 100–101
Context-of-use, 19–36
    dual-task environment, 30–31
    environmental conditions, 31–32
    frequency of use, 33
    International Organization for Standardization, 23–26
    Nielsen, Jakob, 21–26
    Norman, Donald, 22–26
    range of users, 32
    Shackel, Brian, 20–21, 23–26
    Shneiderman, Ben, 23
    training provision, 32–33
    uptake, 33–34
    usability, 20–26
    usability factor specification, 28–29
CPA calculator, 142–143
CPA model, 136–137, 148–150
    advantages, 181
    limitations, 150, 171–172
CPA-predicted task times, 143–145
    data collection, analysis, 144–145
    equipment, 144
    participants, 143
    procedure, 144
Critical path analysis. *See* CPA

D

DALI. *See* Driving activity load index
Direct/indirect input devices, 103–131
    empirical evaluation
        data collection/analysis, 113–114
        IVIS usability, 105

operation types, 107–108
procedure, 112–113
operation types
   eye tracking, 110
   IVIS, 109–110
   University of Southampton, driving simulator, 108–109
   user questionnaires, 110–112
primary driving performance, 114–115
   lateral control, 116–117
   longitudinal control, 115–116
results, 114–130
secondary in-vehicle tasks, 112–113
secondary task performance, 119–122
   secondary task errors, 121–122
   secondary task times, 119–121
subjective measures, 122–126
   driving activity load index, 123–126
   system usability scale, 122–123
usability, 126–129
   graphical user interface, 128
   menu structure usability, 128
   optimisation, 128–129
   rotary controller usability, 127–128
   touch screen usability, 128
visual behaviour, 117–119
Driver needs, 33–47
   in-vehicle tasks, 39
   interaction between driver, IVIS, 38
   multimodal interactions, 44–45
   primary driving tasks, 38–39
   remote controllers, 40–41
   system, 39–41
   task, 38–39
   task-system-user interaction, 45–47
   touch screens, 40–41
   usability prediction, 45
   user, 41–44
   user efficiency, 43
   user enjoyment, 43–44
   user safety, 42
Driving activity load index, 64, 123–126
Driving simulator, University of Southampton, 108–109
Dual-task environment, 30–31
   focal vision, 184–185
   glance behaviour, 159–165
      sequential operations, 163–165
   glance durations, 154–155
   peripheral vision, 184–185
Dual-task IVIS interaction, CPA model, 156–158
   dual-task CPA calculator, 158
   glance behaviour data, 156
   median glance times, 158
   model assumptions, 157
   visual information processing, 157–158
visual IVIS operation initiation, 158
visual mode, 157

E

Efficiency of user, impact of, 43
Empirical evaluation
   data collection/analysis, 113–114
   driving activity load index, 64
   event detection, 62
   IVIS usability, 105
   lateral driving control, 61
   longitudinal driving control, 61
   methods, 60
   objective methods, 60–63
   operation types, 107–108
      eye tracking, 110
      IVIS, 109–110
      University of Southampton, driving simulator, 108–109
      user questionnaires, 110–112
   procedure, 112–113
      secondary in-vehicle tasks, 112–113
   secondary task errors, 63
   secondary task times, 62–63
   subjective methods, 63–64
   system usability scale, 64
   visual behaviour, 61–62
Enjoyment of user, 43–44
Environmental conditions, 31–32
Ergonomics, 12–14
   challenges, 10–12
   human-computer interaction, 12–14
Evaluation of usability, 49–69
   all user groups, 183–184
   analytic evaluation methods
      heuristic analysis, 59
      hierarchical task analysis, 57–58
      layout analysis, 59–60
      multimodal CPA, 58
      systematic human error reduction, prediction approach, 58–59
   analytic methods, 52
   empirical evaluation methods
      driving activity load index, 64
      event detection, 62
      lateral driving control, 61
      longitudinal driving control, 61
      objective methods, 60–63
      secondary task errors, 63
      secondary task times, 62–63
      subjective methods, 63–64
      system usability scale, 64
      visual behaviour, 61–62
   empirical methods, 52
   with experts, 54

Index

information requirements, 50–52
method selection, 49–55
method selection flowchart, 54–55
methods, 55–64
  analytic evaluation methods, 57–60
  empirical evaluation methods, 60
objective measures, 51
persons involved, 53–54
preparation, 49
principles, 50
representing system, tasks, 52–53
representing user, 53
resources, 52–53
subjective measures, 51–52
testing environment, 53
timing, 52
with users, 54
Event detection, 62
Eye tracking, 110

F

Flowchart of method selection, 54–55
Focal vision in dual-task environment, 184–185
Frequency of use, 33
Future predictions, 8–10
Future research, 182–185

G

Glance behaviour, dual-task environment, 159–165
  chunking of visual information, 163
  data, 156
  data analysis, 155–156
  sequential operations, 163–165
  shared glances, 163
  shared visual attention, 159–163
Glances, shared
  dual-task environment glance behaviour, 163
  visual information processing, 179
Graphical user interface, 128
GUI. *See* Graphical user interface

H

HCI. *See* Human-computer interaction
Heuristic analysis, 59, 92–94
  IVIS evaluation, 92
  utility, 92–94
Hierarchical task analysis, 57–58, 77–79
  utility, 79
History of in-vehicle information provision, 1–3
HTA. *See* Hierarchical task analysis
Human-computer interaction
  ergonomics, 12–14

modelling, 133–136
  task times, 134
  techniques, 134–136

I

Identification of operation time, 137–142
In-vehicle comfort control trends, 9
In-vehicle information provision history, 1–3
In-vehicle information systems, 72–74, 109–110
In-vehicle interface analytic evaluation, 71–101
  analytic methods, 72
  data analysis, 76
  equipment, 72–74
    data collection apparatus, 74
  heuristic analysis, 92–94
    utility, 92–94
  hierarchical task analysis, 77–79
  layout analysis, 94–96
    utility, 96
  methods, 72–76
  multimodal CPA, 80–87
    CPA utility, 87
    defining CPA activities, 80–83
    procedure, 74–76
    systematic human error reduction, prediction approach, 87–92
    utility, 92
Indirect/direct input devices, 103–131
  empirical evaluation
    data collection/analysis, 113–114
    IVIS usability, 105
    operation types, 107–108
    procedure, 112–113
  operation types
    eye tracking, 110
    IVIS, 109–110
    University of Southampton, driving simulator, 108–109
    user questionnaires, 110–112
  primary driving performance, 114–115
    lateral control, 116–117
    longitudinal control, 115–116
  procedure, secondary in-vehicle tasks, 112–113
  results, 114–130
  secondary task performance, 119–122
    secondary task errors, 121–122
    secondary task times, 119–121
  subjective measures, 122–126
    driving activity load index, 123–126
    system usability scale, 122–123
  usability, 126–129
    graphical user interface, 128
    menu structure usability, 128
    optimisation, 128–129

rotary controller usability, 127–128
touch screen usability, 128
visual behaviour, 117–119
Industrial context, advantages of evaluation framework in, 180
Information chunking, dual-task environment glance behaviour, 163
Infotainment, 5–7
Instrumentation, 3–5
Interaction between driver, IVIS, 38
International Organization for Standardization, 23–26
ISO. *See* International Organization for Standardization
IVIS. *See* In-vehicle information systems

L

Lateral control, 116–117
Lateral driving control, 61
Layout analysis, 59–60, 94–96
    IVIS evaluation, 94–96
    utility, 96
Longitudinal control, 61, 115–116

M

Manufacturer driver-vehicle interaction strategies, 178
Median glance times, 158
Menu structure usability, 128
Method selection flowchart, usability evaluation, 54–55
Multimethod evaluation framework, 177
Multimodal CPA, 58, 80–87
    CPA utility, 87
    defining CPA activities, 80–83
    IVIS evaluation, 83–87

N

Navigation, 7–8
Needs of driver, 33–47
    in-vehicle tasks, 39
    interaction between driver, IVIS, 38
    primary driving tasks, 38–39
    remote controllers, 40–41
    system, 39–41
    task, 38–39
    task-user-system interaction, 44–45
        multimodal interactions, 44–45
        usability prediction, 45
    touch screens, 40–41
    user, 41–44
    user efficiency, 43
    user enjoyment, 43–44
    user safety, 42
Nielsen, Jakob, 21–26
Norman, Donald, 22–26
Novel contributions of work, 177–179

O

Objective methods, 60–63
Occlusion technique, visual attention, 169
Operation time identification, 137–142
Optimisation, 128–129

P

Peripheral vision in dual-task environment, 184–185
Primary driving performance, 114–115
    lateral control, 116–117
    longitudinal control, 115–116
Primary driving tasks, 38–39

Q

Questionnaires, user, 110–112

R

Range of users, 32
Remote controllers, 40–41
Representing user, usability evaluation, 53
Rotary controller usability, 127–128

S

Safety of user, 42
Screens, touch, 40–41
    usability, 128
Secondary in-vehicle tasks, 112–113
Secondary task errors, 63, 121–122
Secondary task performance, 119–122
    secondary task errors, 121–122
    secondary task times, 119–121
Secondary task times, 62–63, 119–121
Shackel, Brian, 20–21, 23–26
Shared glance CPA model, 165–167
    visual attention, 165–167
Shared glances
    dual-task environment glance behaviour, 163
    visual information processing, 179
Shared visual attention, dual-task environment glance behaviour, 159–163
SHERPA. *See* Systematic human error reduction and prediction approach
Shneiderman, Ben, 23
Simulator, University of Southampton, 108–109
Subjective measures, 122–126
    driving activity load index, 123–126
    system usability scale, 122–123
    usability evaluation, 51–52

Index

SUS. *See* System usability scale
System usability scale, 64, 122–123
Systematic human error reduction and prediction approach, 58–59, 87–92

T

Task-user-system interaction, 44–45
   multimodal interactions, 44–45
   usability prediction, 45
Testing environment, 53
   usability evaluation, 53
Time predictions, 133–152
   CPA, 136–137
   CPA calculator, 142–143
   CPA model, 148–150
   CPA-predicted task times
      data collection, analysis, 144–145
      equipment, 144
      participants, 143
      procedure, 144
      vs. empirical data, 143–145
   extensions to CPA model, 151
   human-computer interaction modelling, 133–136
      task times, 134
      techniques, 134–136
   limitations, CPA model, 150
   operation time identification, 137–142
Timing of usability evaluation, 52
Touch screens, 40–41
   usability, 128
Tracking, 110
Training provision, 32–33

U

University of Southampton, driving simulator, 108–109
Uptake, 33–34
Usability, 20–35, 126–129
   context-of-use, 24
   dual-task environment, 30–31
   environmental conditions, 31–32
   ergonomics, 12–14
   frequency of use, 33
   graphical user interface, 128
   International Organization for Standardization, 23–26
   menu structure usability, 128
   Nielsen, Jakob, 21–26
   Norman, Donald, 22–26
   optimisation, 128–129
   range of users, 32
   rotary controller usability, 127–128
   Shackel, Brian, 20–21, 23–26
   Shneiderman, Ben, 23
   touch screen usability, 128
   training provision, 32–33
   uptake, 33–34
   usability factor specification, 28–29
Usability evaluation, 14–15, 49–69
   analytic evaluation methods
      heuristic analysis, 59
      hierarchical task analysis, 57–58
      layout analysis, 59–60
      multimodal CPA, 58
      systematic human error reduction, prediction approach, 58–59
   analytic methods, 52
   empirical evaluation methods
      driving activity load index, 64
      event detection, 62
      lateral driving control, 61
      longitudinal driving control, 61
      objective methods, 60–63
      secondary task errors, 63
      secondary task times, 62–63
      subjective methods, 63–64
      system usability scale, 64
      visual behaviour, 61–62
   empirical methods, 52
   with experts, 54
   information requirements, 50–52
   method selection, 49–55
   method selection flowchart, 54–55
   methods, 55–64
      analytic evaluation methods, 57–60
      empirical evaluation methods, 60
   objective measures, 51
   persons involved, 53–54
   preparation, 49
   principles, 50
   representing system, tasks, 52–53
   representing user, 53
   requirements, 180
   resources, 52–53
   subjective measures, 51–52
   testing environment, 53
   timing, 52
   with users, 54
Usability factors, specifying, 28–29
User efficiency, 43
User enjoyment, 43–44
User questionnaires, 110–112
User safety, 42

V

Visual attention, 153–173
   CPA model limitations, 171–172
   design implications, 171
   dual-task IVIS interaction, CPA model, 156–158

dual-task CPA calculator, 158
glance behaviour data, 156
median glance times, 158
model assumptions, 157
visual information processing, 157–158
visual IVIS operation initiation, 158
visual mode, 157
glance behaviour, dual-task environment, 159–165
  chunking of visual information, 163
  sequential operations, 163–165
  shared glances, 163
  shared visual attention, 159–163
glance behaviour data analysis, 155–156
implications of CPA findings, 181–182

multidimensionality, 167–168
occlusion technique, 169
road, IVIS glance durations, 169–171
shared glance CPA model, 165–167
visual behaviour, driving, 154–155
  dual-task glance durations, 154–155
visual behaviour theory, driving, 168–169
Visual behaviour, 61–62, 117–119
  driving, 154–155
    dual-task glance durations, 154–155
Visual behaviour theory, driving, 168–169
Visual information chunking, dual-task environment glance behaviour, 163
Visual information processing, 157–158
  shared glances, 179